T0321248

X-RAY PHOTOELECTRON SPECTROSCOPY

X-RAY PHOTOELECTRON SPECTROSCOPY

SPECTROSCOPY

An Introduction to Principles and Practices

PAUL VAN DER HEIDE

WILEY

A JOHN WILEY & SONS, INC., PUBLICATION

Published by John Wiley & Sons, Inc., Hoboken, New Jersey
Published simultaneously in Canada

For general information on our other products and services or for technical support, please contact our Customer Care Department within the United States at (800) 762-2974, outside the United States at (317) 572-3993 or fax (317) 572-4002.

Wiley also publishes its books in a variety of electronic formats. Some content that appears in print may not be available in electronic formats. For more information about Wiley products, visit our web site at www.wiley.com.

Library of Congress Cataloging-in-Publication Data:

Van der Heide, Paul, 1962–
  X-ray photoelectron spectroscopy : an introduction to principles and practices / Paul van der Heide.
       p. cm.
  Includes index.
  ISBN 978-1-118-06253-1 (hardback)
  1. X-ray photoelectron spectroscopy.   I. Title.
  QC454.P48V36 2012
  543'.62–dc23

                                                    2011028942

Printed in the United States of America

10 9 8 7 6 5 4 3 2 1

CONTENTS

FOREWORD

X-ray photoelectron spectroscopy (XPS), ultraviolet photoelectron spectroscopy (UPS), Auger electron spectroscopy (AES), and other related spectroscopies that obtain information about atoms, molecules, compounds, and surfaces by measuring the kinetic energies of electrons ejected from materials have developed rapidly over the past three decades. With their inception in the 1960s at the University of Uppsala, Sweden, under the direction of Professor Kai Siegbahn, these techniques rank as some of the most important advances in surface science and chemical physics in general. Information derivable from these techniques is of vital importance in understanding the electronic structure of solid materials and atoms, molecules, and ions in general.

Dr. Paul van der Heide's book, *X-ray Photoelectron Spectroscopy: An Introduction to Principles and Practices*, treats the phenomenon of photoionization and its consequences by means of modern quantum mechanical methods. The process of ionization, spectroscopic classification of the ionic states, and the subtle spectral vibrational, rotational, vibronic, and spin orbit structure observed in the spectra are discussed in detail. The book serves both a pedagogic need and a research need. It will be of major interest to scientists who use electron spectroscopy in their research and to students majoring in the physical sciences, particularly in the fields of chemistry, physics, and chemical engineering. The book contains a large and valuable collection of references on electron spectroscopy, allowing readers to readily access the original publications. The many examples allow readers to obtain an understanding of some of the subtle and sometimes not so subtle complexities of the photoionization process. It will be particularly useful to senior undergraduates, graduate students majoring in chemistry, physics, and engineering, and scientists and engineers who wish to apply photoelectron techniques in their work.

Department of Chemistry   PROFESSOR JOHN WAYNE RABALAIS
Lamar University
Beaumont, Texas

PREFACE

To know XPS is to know the atom.

X-ray photoelectron spectroscopy (XPS), otherwise known as electron spectroscopy for chemical analysis (ESCA), represents the most heavily used of the electron spectroscopies for defining the elemental composition and/or speciation of the outer 1–10 nm of any solid substrate. This is of importance since it is the solid's surface that defines how the solid is perceived from the outside world. (Note: The composition and/or speciation over the surface region often differs from that of the underlying material) The information content provided by XPS is, however, heavily limited to the untrained eye.

The inspiration for this text arose when teaching the fundamental and practical aspects of XPS. More precisely, this arose on realizing how the collective works and experiences could be used in preparing a text that would further facilitate this transfer of knowledge. The underlying premise used in putting together this book was *easily attainable answers to all of the questions asked over the years*. For example, "Why is XPS so effective in analyzing all the elements except for hydrogen or helium, when all it is doing is kicking out electrons?"[1] (Hydrogen and helium have electrons too.)

In adhering to this premise, all aspects associated with XPS are introduced from a practitioners perspective; all sections are prepared such that they can be read independently of each other; all equations are presented using the most commonly used units; and all fundamental aspects are discussed using classical analogies, where possible, over the more correct quantum mechanics descriptions. Note: Although XPS has its foundations in quantum mechanics, knowledge of this is not needed to understand XPS.

The ordering of the text is as follows: Since XPS derives information on the elemental composition and speciation by probing the electronic structure of the system in question (this is first covered in the introduction), an overview of the electronic structure of atoms is presented. Following this, the practical aspects of XPS from basic analysis procedures to instrumentation are covered. An overview of spectral interpretation is then presented followed by several illustrative case studies. Lastly, various complementary techniques and related concepts are covered in the Appendix.

In preparing this text, it also became apparent that this could be used as an effective stepping stone to some of the more extensive publications available, some of which contain a wealth of useful information that can otherwise be inaccessible to the uninitiated. What separates this text from other introductory texts is the greater emphasis placed on spectral understanding/interpretation. This is considered important since therein lies the ability of XPS to define speciation.

PAUL A.W. VAN DER HEIDE

---

[1] Hydrogen and helium within solids are not detectable by XPS because their photoelectron cross sections (yields) are below XPS detection limits. This arises from a combination of effects, namely, that

(a) Photoelectron cross-sections (these govern intensities) from stationary states of a specific principal quantum number decrease with decreasing atomic number.

(b) XPS is not highly sensitive to valence electrons (hydrogen and helium only have valence electrons, while XPS analysis is optimized toward core electrons).

In addition, since all of their electrons are within the valence region, there is nothing characteristic about their binding energies ($B.E.$s); that is, these participate in bonding.

Note: The increased sensitivity of ultraviolet photoelectron spectroscopy (UPS) to valence electrons can allow for the detection of hydrogen and helium in the gas phase.

ACKNOWLEDGMENTS

Although this book has profited from many people, there are three in particular whose names deserve mention. These are Professor John Wayne Rabalais, Dr. David Surman, and Kim van der Heide for the invaluable assistance, encouragement, and/or comments provided in preparing this text. Thank you.

P.V.D.H.

LIST OF CONSTANTS

Boltzmann constant	k_B	1.381×10^{-23} J/K
or in units of eV/K		8.616×10^{-5} eV/K
Charge of electron	e	1.602×10^{-19} C
Mass of electron	m_e	9.109×10^{-31} g
Mass of neutron	m_n	1.675×10^{-27} g
Mass of proton	m_p	1.673×10^{-27} g
Planck constant or in	$h = \hbar \cdot 2\pi$	6.626×10^{-34} J·s
units of eV·s		4.136×10^{-15} eV·s
Speed of light	c	2.98×10^{8} m/s

CHAPTER 1

INTRODUCTION

1.1 SURFACE ANALYSIS

We interact with our surroundings through our five senses: taste, touch, smell, hearing, and sight. The first three require signals to be transferred through some form of interface (our skin, taste buds, and/or smell receptors). An interface represents two distinct forms of matter that are in direct contact with each other. These may also be in the same or different phases (gas, liquid, or solid). How these distinct forms of matter interact depends on the physical properties of the layers in contact.

The physical properties of matter are defined in one form or another by the elements present (the types of atoms) and how these elements bond to each other (these are covered further in Section 2.1). The latter is referred to as *speciation.*

An example of speciation is aluminum (spelt aluminium outside the United States) present in the metal form versus aluminum present in the oxide form (Al_2O_3). In these cases, aluminum exists in two different oxidation states (Al^0 vs. Al^{3+}) with highly diverse properties. As an example, the former can be highly explosive when the powder form is dispersed in an oxidizing environment (this acted as a booster

X-ray Photoelectron Spectroscopy: An Introduction to Principles and Practices,
First Edition. Paul van der Heide.
© 2012 John Wiley & Sons, Inc. Published 2012 by John Wiley & Sons, Inc.

rocket propellant for the space shuttle when mixed with ammonium perchlorate), while the latter is extremely inert (this is the primary form aluminum exists within the earth's crust).

Aluminum foil (the common household product) is primarily metallic. This, however, is completely inert to the environment (air under standard temperature and pressure) since it is covered by a thin oxide layer that naturally reforms when compromised. This layer is otherwise referred to as a *passivating oxide*. Note: Aluminum metal does not occur naturally. This is a man-made product whose cost of manufacture has decreased dramatically over the last 200 years. Indeed, aluminum metal was once considered more precious than gold, and it is reputed that Napoleon III honored his favored guests by providing them with aluminum cutlery with the less favored guests being provided with gold cutlery.

Like aluminum foil, most forms of matter present in the solid or liquid phase exhibit a surface layer that is different from that of the underlying material. This difference could be chemical (composition and/or speciation), structural (differences in bond angles or bond lengths), or both. How a material is perceived by the outside world thus depends on the form of the outer layer (cf. an object's *skin* or *shell*). The underlying material is referred to as the bulk throughout the remainder of this text. Also, gases are not considered due to their high permeability, a fact resulting from a lack of intermolecular forces and the high velocity of the constituents (N_2 and O_2 in air travel on average close to 500 m/s, with any subsequent collisions defining pressure).

Reasons as to why the physical properties of a solid or liquid surface may vary from the underlying bulk can be subdivided into two categories, these being

(a) *External Forces (i.e., Adsorption and/or Corrosion of the Outer Surface)*. Pieces of aluminum or silicon are two examples in which a stable oxide (passivation layer) is formed on the outer surface that is only a few atomic layers thick (~1 nm). Note: Air is a reactive medium. Indeed, water vapor catalyzes the adsorption of CO_2 on many metallic surfaces (both water vapor and CO_2 are present in air), and so forth.

(b) *Internal Forces (i.e., Those Relayed through Surface Free Energy)*. These are introduced by the abrupt termination of any long-range atomic structure present and can induce such effects as elemental segregation, structural modification (relaxation and/or reconstruction), and so on. This too may only influence the outer few atomic layers.

Some of the physical properties (listed in alphabetical order) that can be affected as a result of these modifications (notable overlaps existing between these) include

(a) Adhesion
(b) Adsorption
(c) Biocompatibility
(d) Corrosion
(e) Desorption
(f) Interfacial electrical properties
(g) Reactivity inclusive of heterogeneous catalysis
(h) Texture
(i) Visible properties
(j) Wear and tear (also referred to as tribology)
(k) Wetability, and so on

If the surface composition and speciation can be characterized, the manner in which the respective solid or liquid interacts with its surroundings can more effectively be understood. This, then, introduces the possibility of modifying (tailoring) these properties as desired. From a technological standpoint, this has resulted in numerous breakthroughs in almost every area in which surfaces play a role. Some areas (listed in alphabetical order) in which such modifications have been applied include

(a) Adhesion research
(b) Automotive industry
(c) Biosciences
(d) Electronics industry
(e) Energy industry
(f) Medical industry
(g) Metallurgy industry inclusive of corrosion prevention
(h) Pharmaceutical industry
(i) Polymer research, and so on

Indeed, many of these breakthroughs have resulted from the tailoring of specific surface properties and/or the formulation of new materials that did not previously exist in nature. Like aluminum foil, these are all man-made with examples ranging from the development of plastics to synthesis of superconducting oxides, and so on.

A solid or liquid's surface can be defined in several different ways. The more obvious definition is that *a surface represents the outer or topmost boundary of an object.* When getting down to the atomic level, however, the term boundary loses its definition since the orbits of bound electrons are highly diffuse. An alternative definition would then be that *a surface is the region that dictates how the solid or liquid interacts with its surroundings.* Applying this definition, a surface can span as little as one atomic layer (0.1–0.3 nm) to many hundreds of atomic layers (100 nm or more) depending on the material, its environment, and the property of interest.

To put these dimensions into perspective, consider a strand of human hair. This measures between 50 and 100 μm (0.05–0.1 mm) in diameter. The atoms making up the outer surface are of the order of 0.2 nm in diameter. This cannot be viewed even under the most specialized optical microscope (typical magnification is up to ~300×) since the spatial resolution is diffraction limited to values slightly less than 1 μm (see Appendix E). The magnification needed (~30,000,000×) can only be reached using a very limited number of techniques, with the most common being transmission electron microscopy (TEM). These concepts are illustrated in Figure 1.1.

TEM being a microscopy, however, only reveals the physical structure of the object in question. To reveal the chemistry requires spectroscopy or spectrometry (the original difference in terminology is discussed in Appendix F). Although a plethora of spectroscopies and spectrometries exists, few are capable of providing the chemistry active over the outermost surface, that is, that within the outermost 10 nm of a solid. Of the few available, X-ray photoelectron spectroscopy (XPS), also referred to as electron spectroscopy for chemical analysis (ESCA),

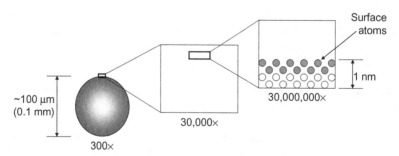

Figure 1.1. Pictorial illustration of the cross section of a strand of human hair at the various magnifications listed whose surface may have been modified to add or remove specific properties, that is, dryness, oiliness, cleanliness, and sheen, through, for example, the application of a specific shampoo.

has over the last several decades become the most popular. Some comparable/complementary microanalytical techniques are discussed in Appendices F and G.

1.2 XPS/ESCA FOR SURFACE ANALYSIS

XPS, also referred to as ESCA, represents the most heavily used of the electron spectroscopies (those that sample the electron emissions) for defining the elemental composition of a solid's outer surface (within the first 10 nm). The acronym XPS will be used henceforth in this text since this more precisely describes the technique. The acronym ESCA was initially suggested by Kai Siegbahn when realizing that speciation could be derived from the photoelectron and Auger electron emissions alone.

The popularity of XPS stems from its ability to

(a) Identify and quantify the elemental composition of the outer 10 nm or less of any solid surface with all elements from Li–U detectable. Note: This is on the assumption that the element of interest exists at >0.05 atomic % (H and He are not detectable due to their extremely low photoelectron cross sections and the fact that XPS is optimized to analyze core electrons).

(b) Reveal the chemical environment where the respective element exists in, that is, the speciation of the respective elements observed.

(c) Obtain the above information with relative ease and minimal sample preparation.

Aside from ultraviolet photoelectron spectroscopy (UPS), which can be thought of as an extension of XPS since this measures the valence band photoelectrons, Auger electron spectroscopy (AES) is the most closely related technique to XPS in that it displays a similar surface specificity while being sensitive to the same elements (Li–U). Its strength lies in its improved spatial resolution, albeit at the cost of sensitivity (for further comparisons of related techniques, see Appendix F).

Wavelength-dispersive X-ray analysis (WDX) and energy-dispersive X-ray analysis (EDS or EDX) are also effective for defining the elemental composition of solids. Indeed, when combined with scanning electron microscopy (SEM), these are more popular than XPS, with the moniker electron probe microanalysis (EPMA) often used. These, however, are not considered true surface analytical techniques, at least not in the strict sense, since they provide average elemental concentrations over a depth that extends ~1 μm or more below the surface.

This difference in depth is important since nearly all the surface chemistry that takes place between different forms of matter is dictated by the chemical composition, speciation, and/or electronic structure present over the outer few atomic layers of the respective solid. These can differ substantially from those noted 1 nm or more below the surface. Note: A surface film of 1 nm equates to three to four atomic layers. Thus, the examination of aluminum foil via EDX reveals spectra heavily dominated by aluminum peaks. This would not reveal the presence of a surface oxide.

1.3 HISTORICAL PERSPECTIVE

Historically, XPS can be traced back to the 1880s whereupon Heinrich Hertz noted that electrically isolated metallic objects held under vacuum exhibited an enhanced ability to spark when exposed to light (Hertz, 1887). This effect, termed the *Hertz effect*, also allowed the derivation of the ratio of Plank's constant over electronic charge (h/e) and the work function (ϕ) of the respective metal object when altering the energy of the irradiation source (frequency, wavelength, and energy are all related).

In 1905, Albert Einstein explained this effect as arising from the transfer of energy from photons (in the form of light) to electrons bound within the atoms making up the respective metallic objects. In other words, he showed that this induced electron emission from metallic objects if the energy transfer was greater than the energy that binds the electron to the respective metal atom/solid (Einstein, 1905). For this and the introduction of the concept of the *photon* (a package of energy with zero rest mass), Einstein was awarded the 1921 Nobel Prize in Physics.

The above-mentioned spark can thus be understood as resulting from the net positive charge that builds on photoelectron and Auger electron emission from electrically isolated objects. Note: Photoelectron emission is also accompanied by Auger electron emission or fluorescence (emission of photons). Auger emission is named after Pierre Auger (1925) but was first reported by Lisa Meitner (Meitner, 1922), while fluorescence was named by George Stokes, who was first responsible for bringing about a physical understanding underlying this phenomena (Stokes, 1852).

The capabilities of XPS were, however, not fully recognized until Kai Siegbahn and his coworkers constructed an instrument capable of analyzing core photoelectron emissions to a sufficiently high energy resolution to allow speciation analysis to be carried out (Siegbahn, 1967, 1970). For this, Kai Siegbahn was awarded the 1981 Nobel Prize in Physics. Following Siegbahn's initial success came a rapid succession of studies and instruments (for a brief synopsis, see Shirley and Fadley, 2004) that resulted in a firm understanding of XPS. The primary reason for this relatively recent development can be traced back to the inability to attain the necessary vacuum conditions required when analyzing such surface regions (the need for vacuum is discussed in further detail in Section 3.1.1). Note: The requirement for vacuum generally limits the application of this technique to the analysis of solid surfaces. Gases and liquids can be analyzed but only when using highly specific instrumentation and/or sample preparation procedures.

1.4 PHYSICAL BASIS OF XPS

Photoelectron production in its simplest form describes a single step process in which an electron initially bound to an atom/ion is ejected by a photon. Since photons are a massless (zero rest mass), chargeless package of energy, these are annihilated during photon–electron interaction with complete energy transfer occurring. If this energy is sufficient, it will result in the emission of the electron from the atom/ion as well as the solid. The kinetic energy ($K.E.$) that remains on the emitted electron is the quantity measured. This is useful since this is of a discrete nature and is a function of the electron binding energy ($B.E.$), which, in turn, is element and environment specific.

A schematic example of the photoelectron emission process from oxygen present within a silicon wafer bearing a native oxide is shown on the left in Figure 1.2a. As covered in Section 2.1.2.2, photoelectron peaks are described using spectroscopic notation. To the right of Figure 1.2a is shown one of the two primary de-excitation processes that follow photoelectron emission, that is, the Auger process. The other process, termed fluorescence, results in photon emission. These are described using X-ray notation. (Note: This can be confusing since the same levels are described.) Since Auger de-excitation also results in electron emission, peaks from both photoelectrons and Auger electrons are observed in XPS spectra. Further discussion on X-ray-induced Auger emission is covered in Section 5.1.1.3.2.4.

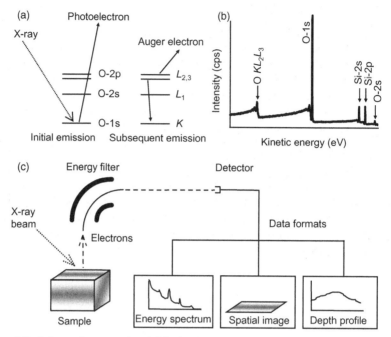

Figure 1.2. Schematic example of (a) the photoelectron process (shown on the left) and a subsequent Auger de-excitation process (shown on the right) with the various electronic energy levels (stationary states) portrayed using either spectroscopic notation (photoelectron peaks) or X-ray notation (Auger electron peaks), (b) XPS spectra collected from a silicon wafer bearing a surface oxide as analyzed under Mg-$K\alpha$ irradiation (as described in the text, this contains peaks from both photoelectron emissions and Auger electron emissions), and (c) the basic components of an XPS instrument along with the data formats that can be implemented. Further discussion on the instrumentation required along with the acquisition of energy spectra along with spatial images is covered in Chapter 3, while depth profiling is covered in Section 4.3.1.2.

In Figure 1.2b, a typical low-resolution spectrum collected from a silicon wafer is shown. This is plotted in intensity (I) versus $K.E.$ (the energy the electron emissions attain on departing the sample). Evident in this spectrum are photoelectrons from electronic levels accessible to the X-ray source used, that is, the O-1s, O-2s, Si-2s, and Si-2p levels (here, the Si-2p$_{1/2}$ and Si-2p$_{3/2}$ contributions overlap) as well as Auger electron emissions resulting from the filling of the O-1s core hole (K level). The latter are, however, described using X-ray notation (see Section 2.1.2.2), that is, as O-KLL emissions or some specific contribution derivative thereof (in this case the KL_2L_3 emissions).

O-KLL emissions arise from the filling of the K level core hole produced on photoelectron emission by an electron from some L level, with the energy difference between these two levels carried away in the emission of a third electron, also from some L level. The most intense of the peaks actually arises from KL_2L_3 and KL_3L_2 transitions collectively referred to as $KL_{2,3}L_{2,3}$ emissions. The remaining peaks arise from KL_1L_1, KL_1L_2, and KL_1L_3 emissions with the latter two collectively referred to as $KL_1L_{2,3}$.

In Figure 1.2c is shown a schematic example of an XPS instrument, along with the three most common means of relaying the data, namely,

(a) Energy distributions of any electron emissions falling within some predefined energy range

(b) Spatial distributions of specific electron emissions noted across a surface (this allows the elemental or speciation distributions to be mapped)

(c) Depth distributions of specific electron emissions to some predefined depth (this can extend from less than 10 nm to several micrometers)

Analysis is usually carried out by first collecting energy spectra over all accessible energies and then concentrating on particular photoelectron signals. This ensures that all elements are accounted for during quantification and that the data are collected in a time-effective manner.

Although $K.E._{\text{XPS}}$ is the quantity recorded in XPS, it is the derived $B.E._{\text{XPS}}$ that is used to construct the energy spectrum. Note: The XPS subscript is applied henceforth to denote the fact that the value obtained is not exactly equal to that expected in a ground-state atom; that is, the introduction of a core hole during photoemission effectively alters $B.E.$ values from that exhibited by a ground-state atom/ion, albeit by a small amount. This effect, referred to as a *final state effect*, is discussed in Section 5.1.1.3.

The $B.E._{\text{XPS}}$ derived is used to construct a spectrum since the $K.E._{\text{XPS}}$ is dependent on the X-ray energy, whereas the $B.E._{\text{XPS}}$ is not. Values of $K.E._{\text{XPS}}$, $B.E._{\text{XPS}}$, and the initiating photon energy (E_{ph}) are related through the expression (Einstein, 1905)

$$K.E._{\text{XPS}} = E_{\text{ph}} - \phi_{\text{XPS}} - B.E._{\text{XPS}}, \tag{1.1}$$

where ϕ_{XPS} is the work function of the instrument, not the sample. This is included since it represents the minimum energy necessary to remove an electron from the instrument on the assumption that a conductive sample in physical contact with the instrument is analyzed (the use of this as opposed to that of the sample is discussed in Section 4.1.3).

Note: Equation 1.1 does not apply to Auger emissions noted in XPS spectra. This is realized since Auger electron energies are not directly related to the incoming photon energy (E_{ph}); rather, these represent the difference between two energy levels once electronic perturbation effects are accounted for (Auger emission is discussed further in Section 5.1.1.3.2.4). Applying Equation 1.1 to spectra obtained under different X-ray energies thus yields different values for Auger electrons. Indeed, this can be useful when there exists confusion as to whether an observed peak is a photoelectron peak or an Auger electron peak; that is, this effectively shifts the Auger peaks along the $B.E._{XPS}$ scale as discussed in Section 4.1.2.

Representative $B.E._{XPS}$ values for all the elements can be found in Appendix B.

1.5 SENSITIVITY AND SPECIFICITY OF XPS

Two parameters that describe the ability of XPS to identify and quantify the elemental composition and speciation present over the outer 10 nm or less of any solid surface, on the assumption that the element of interest exists at >0.05 atomic %, are

(a) Surface specificity or the ability to separate the signal from the surface region relative to that of the underlying region
(b) Sensitivity or the ability to detect the signal of interest given the constraint of the reduced volume from which the signal emanates

Surface specificity arises from the limited flight path an electron has within a solid before it loses some fraction of its energy (this is generally less than 10 nm as discussed in Section 4.2.2.1). Note: X-rays can penetrate micrometers below the surface. If energy is lost, the signal will disappear within the spectral background (see Section 4.2.3). This occurs for almost all photoelectrons produced from atoms/ions situated at some depth greater than ~10 nm below the surface. Hence, the discrete signals that remain (those that result in the spectral peaks observed) are from the surface region alone. The presence of an adsorbed surface layer of some thickness will thus act to quench, to some degree, all signals from the underlying substrate. This otherwise reduces the sensitivity of XPS to these elements. These concepts are illustrated in Figure 1.3.

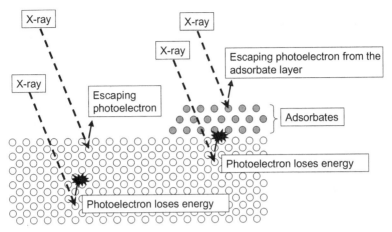

Figure 1.3. Pictorial illustration of photoelectron emission (those with a $K.E._{XPS}$ of ~100 eV) from a solid material (represented by the hollow circles) and the surface adsorbate layer (represented by the gray circles).

Sensitivity is primarily a function the photoelectron cross section and the spectral background level (these are discussed further in Sections 4.2.1 and 4.2.3, respectively). The photoelectron cross section describes the yield of electrons produced as a function of the impacting photon energy. Indeed, the low photoelectron cross sections from H and He under X-ray irradiation are the primary reason why these elements are not detectable when present within solids. The ambient pressure, or more precisely the vacuum under which the analysis is carried out, can also affect sensitivity since this controls the density of molecules in the gas phase and, thus, the flight path of any photoelectrons emanating from the surface. In other words, this acts to restrict the passage of electrons from the sample to the detector. Contaminant overlays also form within analysis timescales if the pressure under which the analysis is being carried out is too high. Pressure and its impact in XPS are discussed in greater detail in Section 3.1.1.1.

1.6 SUMMARY

The properties of a solid or liquid as viewed from the outside world are primarily dictated by the physical properties of the outermost layer. Indeed, this region, termed the surface, can differ in composition and/or structure from that of the underlying bulk material. As a result, much effort has been expended in the understanding of this region. To complicate matters, this region can span as little as a few atomic layers (1 nm or less) to many hundreds of atomic layers (up to 100 nm).

With the advent of the technology necessary to produce the vacuum conditions required to effectively analyze a solid's surface came the ability to measure the surface chemistry. This culminated in the 1960s in the development of a technique that has since become the most popular and heavily used of the microanalytical techniques for examining the chemistry active on or within solid surfaces. This technique, termed XPS, is now the mainstay of almost all surface analysis labs worldwide, whether in academic or industrial settings. Attributes that have lent to the popularity of XPS include

(a) Elemental identification and quantification of any element from Li to U

(b) Sensitivity (concentrations down to 0.1 atomic %)

(c) Surface specificity (less than 10 nm)

(d) Ease of analysis (minimal sample preparation is required)

As first explained by Einstein, XPS derives this information by directing an energetic photon beam (X-rays) to induce the emission of core-level electrons. The energy of the electron emissions is then measured since this provides insight into the specific type of atoms/ions the electrons emanated from (the elements), the amount or ratios of the respective atoms/ions (the composition of the volume analyzed), and, in many cases, the manner in which the element was bound (the speciation of the atoms/ions within the volume analyzed), all with relative ease.

CHAPTER 2

ATOMS, IONS, AND THEIR ELECTRONIC STRUCTURE

2.1 ATOMS, IONS, AND MATTER

All matter, whether present in the gas, liquid, or solid states, comprises a grouping of atoms and/or ions. The properties exhibited by these groupings of atoms and/or ions are dictated by the type of atoms and/or ions present and by the form in which they are bound, that is, their speciation.

Atoms/ions are comprised of a highly dense nucleus around which electrons orbit, much like the planets orbit around the sun. The nucleus is comprised of protons and neutrons made up of elementary particles called quarks (electrons are also elementary particles). Protons have one unit of positive charge; neutrons are neutral; and electrons have one unit of negative charge. Atoms are neutral in charge since the number of protons equates to the number of electrons. Ions are atoms that have gained or lost one or more electrons. Ions can be formed through bonding or if the atom experiences a sufficiently energetic collision such that the energy transferred results in the emission of adsorption of an electron.

X-ray photoelectron spectroscopy (XPS) derives information on the elemental composition and speciation of matter (most commonly in the

X-ray Photoelectron Spectroscopy: An Introduction to Principles and Practices,
First Edition. Paul van der Heide.
© 2012 John Wiley & Sons, Inc. Published 2012 by John Wiley & Sons, Inc.

solid state) by assessing the electronic structure of the atoms and/or ions residing within the surface region of the sample (matter) being analyzed.

2.1.1 Atomic Structure

An atom represents the smallest indivisible unit of matter; that is, this cannot be broken down further unless exposed to an exorbitant amount of energy. Atoms range in size from 62 pm (1 pm is 10^{-12} m) for He to 520 pm for Cs. Atoms are made up of *protons*, *neutrons*, and *electrons*. The protons and neutrons reside in the highly dense *nucleus* around which the electrons orbit. The dimension of the nucleus is of the order of femtometers (1 fm is 10^{-15} m).

Free atoms (unbound) prefer to be in a state in which they have zero charge. This will arise if the number of protons and electrons within the atom is equal (recall protons have a unit charge of +1; electrons have a unit charge of –1; and neutrons have no charge). The number of electrons gained or lost is represented by a superscript following the element symbol. As an example, He^{2+} is a helium atom (alpha particle) that has lost its two electrons. If neutral, no superscript is used.

The *elements* listed in the periodic table (see Appendix A) are atoms with different chemical properties. These properties define how each element reacts with other elements, and the types of molecules or solids they form. In short, the properties of the elements, and thus the molecules or solids they form, are a direct function of the number and distribution of electrons orbiting the respective nuclei (this dictates the type of bond formed). Since the number of electrons is defined by the number of protons in the respective atoms' nucleus, the number of protons defines the element. For example, H has one proton, He has two protons, and Li has three protons. This number is also referred to as *atomic number* (Z), which is expressed in the periodic table above the element symbol. The number of neutrons defines the isotope. The *atomic weight* typically listed below the element symbol in the periodic table is the average atomic mass for all the isotopes of the respective element.

The word *atom* comes from the Greek work *atomus*, which means uncuttable. This concept was first introduced by the Greek philosopher Democritus. Sufficient evidence for this concept was not produced, however, until the early 19th century, for which Dalton is primarily accredited for (Roscoe, 1895). Indeed, the SI accepted unit for atomic mass is dalton, even though the unified atomic mass unit, is more commonly used (these are all equal to each other). The official SI unit for mass is gram. This, however, is much too large as shown in Table 2.1.

TABLE 2.1   Rest Mass and Charge of Free Protons, Neutrons, and Electrons

Particle	Mass (g)/Symbol	Charge (C)
Proton	1.6726×10^{-24} g/m_p	1.6022×10^{-19}
Neutron	1.6749×10^{-24} g/m_n	0
Electron	9.1094×10^{-28} g/m_e	1.6022×10^{-19}

Note: 1 u equals 1.6605×10^{-24} g, and one unit of charge (q) equates to 1.6022×10^{-19} C.

The structure of the atom was revealed in 1909 by Ernst Rutherford, who is also generally credited for the discovery of the proton (Rutherford, 1911). The neutron was not discovered until 1932 (Chadwick, 1932). The electron was discovered by J.J. Thomson in 1897 (Thomson, 1897). Interestingly, Chadwick was a student of Rutherford who, in turn, was a student of J.J. Thomson.

2.1.2   Electronic Structure

The electronic structure describes the energies and spatial distribution of all bound electrons within a respective atom, ion, molecule, or solid. Electrons are bound to atoms/ions through *electromagnetic force* (protons in the nuclei have a positive charge and spin, while electrons have a negative charge and spin). The extent of the attraction is defined as the electron *binding energy* (*B.E.*), which is a function of

(a) The number of protons in the nucleus (this also defines the element; i.e., H has one proton, He has two protons, and Li has two)

(b) The distance between the core electrons and their nuclei (defined by the stationary state the electron resides in)

(c) The density of electrons around the respective atom, ion, molecule, or solid (influenced by the type of bonding that occurs)

(d) The electron–electron interactions present (electrons repel each other as well as shield electrons of lesser *B.E.* from the nuclei)

As specified by quantum mechanics, each bound electron has its own set of *quantum numbers*. These numbers, discussed in Section 2.1.2.1, reveal which *stationary state* the respective electron resides in. The states in which these electrons reside are described using either *spectroscopic notation* or *X-ray notation*. These are described further in Section 2.1.2.2. Both describe the same thing and hence can be viewed as different languages (one developed by chemists and the other by physicists).

Electrons can also move between various stationary states within the same atom/ion or neighboring atoms/ions. These transitions can be described using an extension of either spectroscopic notation, or X-ray notation, or some derivative thereof. Indeed, when photons (X-rays) are produced, Seigbahn notation is often applied. These are discussed further in Section 2.1.2.3.

These states, commonly referred to as stationary states, describe discrete energy levels in which an electron can reside in when bound to an atom/ion. These energies are specific to each element and, to a lesser extent, the chemical environment the element exists in (speciation describes the effect of the chemical environment on the respective element). Further discussion on stationary states is presented in Section 2.1.2.4.

The chemical environment plays a role since atoms/ions bond to each other through the interaction of their *valence electrons* (their outermost electrons), again via the electromagnetic force. This, in turn, alters the energies of all inner electrons, referred to as the *core electrons*, according to the type of bonding active as this affects the resulting electron density. Electrons in stationary states with nonzero angular momentum (angular momentum is a quantum number) also experience what is referred to as *spin orbit splitting*. This causes a splitting of the respective energy level (stationary state) as discussed further in Section 2.1.2.5.

Due to the different energy references used, apparent variations in core electron *B.E.* values are also noted for atoms/ions in different phases; that is, energies of electrons bound to atoms/ions in the gas phase are referenced to the *vacuum level* (E_{vac}), whereas electrons bound to atoms/ions in the solid phase are referenced to the *Fermi edge* (E_F). The solids *work function* (ϕ) defines the difference. In other words, this represents the minimum energy required to remove an electron from a solid, which is analogous to *ionization potential* (I) being the minimum energy to remove an electron from a free atom/ion. The aforementioned concepts are illustrated pictorially for an arsenic atom in Figure 2.1.

2.1.2.1 Quantum Numbers

Although quantum mechanics is a mathematical interpretation, its implications have great physical significance (Griffiths, 2004). Indeed, the energy levels implied are detected by the following electron spectroscopies: XPS, Auger electron spectroscopy (AES), electron energy loss spectroscopy (EELS), and so on. Further discussion on various quantum mechanics-based calculations can be found in Appendix C.

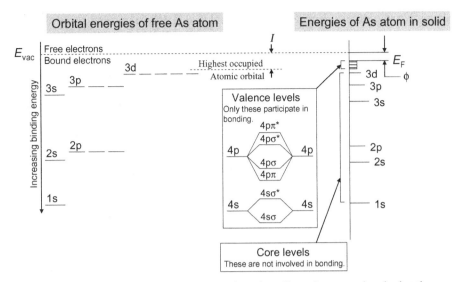

Figure 2.1. A schematic diagram illustrating the allowed energy levels (stationary states) for electrons bound to a free arsenic atom (left) and the allowed energy levels for electrons bound to arsenic when present within a solid (right). Bonding results in the formation of molecular orbitals in which only the valence electrons take part (those in the 4s and 4p levels). Spectroscopic notation is used and spin orbit splitting is not considered.

In quantum mechanics, each electron is assigned a *principal quantum number* (n). This defines the energy and spatial extent of the electron's location around the nucleus. It is an integer number starting from 1 for the most tightly bound electrons (those closest to the nucleus). This number increases in ascending order with decreasing *B.E.* and can contain one or more substates depending on the value of n. Substates are defined by the shape and orientation of the electron's distribution around the nucleus.

The shape of the electron's spatial distribution around the nucleus is represented by the *angular momentum quantum number* (l). This can have integer values from 0 to $n - 1$. For example, electrons in the $n = 3$ state can exist in the $l = 0$, $l = 1$, and $l = 2$ substates. Electrons in $l = 0$ states exhibit spherical distributions around the nucleus. Electrons in $l > 0$ states exhibit l dependent nonspherical distributions around the respective nucleus.

The orientation of an electron's spatial distribution around the nucleus is represented by the electron's *magnetic quantum number* (m_l). This can have integer values that extend from $-l$ to 0 to $+l$. These levels can only contain two electrons, both of which must have opposing spins.

Since every electron has a spin (s) equal to 1/2, these spin states, defined by the spin quantum number (m_s), take on values of +1/2 and −1/2.

The quantum numbers n, l, m_l, and m_s represent the basis of all other quantum numbers. For example, L, M, and S define the sum of l, m_l, or m_s, respectively. Likewise, j and J are defined via the vectorial addition of $l + m_s$ (also portrayed as $l \pm s$) and $L + S$ (also portrayed as the sum of j), respectively. Thus, both j and J represent the total angular momentum, with j defining that of an individual electron and J that of the entire atom, ion, or molecule.

Note: No two electrons bound to the same atom/ion can have the same set of quantum numbers. This is otherwise referred to as the *Pauli exclusion principle*.

2.1.2.2 *Stationary-State Notation*

Two types of nomenclature are used to describe the various stationary states present in an atom/ion, irrespective of whether they are occupied or not. These are

(a) Spectroscopic notation
(b) X-ray notation

Both describe the same thing, and both are directly related to the quantum numbers of a bound electron, if present. These interrelations are illustrated for electrons with quantum numbers up to $n = 4$ and $l = 3$ in Figure 2.2.

The difference between the nomenclatures can be traced back to the communities that developed these; that is, spectroscopic notation was developed by chemists, while X-ray notation was developed by physicists. As a result, XPS, which was developed by chemists, uses spectroscopic notation, while AES, X-ray fluorescence (XRF), and EELS, which were developed by physicists, all use X-ray notation.

In short, spectroscopic notation uses a specific integer followed by a specific letter to define a particular stationary state. The number used is the same as the principal quantum number of the electron that resides in this level. The letter, on the other hand, relates to the angular quantum number, for example,

(a) An s orbital (*S*harp) describes the stationary state occupied by electrons with the angular momentum quantum number $l = 0$.
(b) A p orbital (*P*rinciple) describes the stationary state occupied by electrons with the angular momentum quantum number $l = 1$.

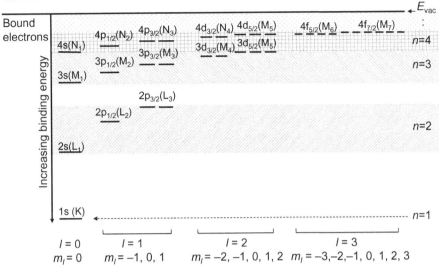

Figure 2.2. Schematic representation (not to scale) of the available stationary states within all ground-state atoms/ions, along with the various quantum numbers an electron would have if present when in these states (these are listed to the right and along the bottom of this figure). The spectroscopic notation describing these levels is shown above the respective stationary state (depicted by the short horizontal lines), while X-ray notation is shown above the level but in brackets. *Note:* These start to merge with increasing n, that is, as E_F is approached.

(c) A d orbital (*Diffuse*) describes the stationary state occupied by electrons with the angular momentum quantum number $l = 2$.

(d) An f orbital (*Fundamental*) describes the stationary state occupied by electrons with the angular momentum quantum number $l = 3$.

A variation in the energy of a stationary state containing electrons with the same n and l quantum numbers can occur if spin orbit splitting is in effect (this is discussed in Section 2.1.2.5). The energy levels resulting from this splitting are signified by the quantum number j presented in subscript form following the principle and angular momentum designation. For example, the 2p level has two energetically distinct levels arising from spin orbit splitting that are referred to as the $2p_{1/2}$ and $2p_{3/2}$ levels. Note all p, d, and f orbitals split into doublets described by the respective j values.

X-ray notation in its simplest form uses capital letters K, L, M, N, O, and so on, to denote the principal quantum number, starting with K for electrons with $n = 1$ and proceeding alphabetically to higher principal quantum numbers. Since these are core levels, they are sometimes more simplistically represented using the letter C. Levels within the valence region, that is, those close to E_{vac}, can become indistinguishable in energy. As a result, these are often assigned the letter V (for valence).

Sublevels within the same principal quantum number are designated by an integer value appearing as a subscript following the capital letter. The value of 1 is used for the level containing electrons most tightly bound to the atom/ion, with subsequent sublevels denoted in ascending order. Thus, the 2s level is represented as L_1, and the $2p_{1/2}$ and $2p_{3/2}$ levels are referred to as the L_2 and L_3 levels, respectively.

2.1.2.3 Stationary-State Transition Notation
Spectroscopic notation, X-ray notation, or some derivative thereof is also used to describe transitions between different stationary states.

For example, the production of a 1s core hole in an aluminum atom (Step 1) followed by the transition of a $2p_{3/2}$ electron to fill this core hole (Step 2) can be described using

Step 1: $1s^2 2s^2 2p^6 3s^2 3p^1 \rightarrow 1s^1 2s^2 2p^6 3s^2 3p^1$
Step 2: $1s^1 2s^2 2p^6 3s^2 3p^1 \rightarrow 1s^2 2s^2 2p^5 3s^2 3p^1$

To simplify matters, only the levels affected may be presented; that is,

Step 1: $1s^2 2p^6 \rightarrow 1s^1 2p^6$
Step 2: $1s^1 2p^6 \rightarrow 1s^2 2p^5$

If X-ray notation were to be used, Step 2 would be represented as a KL_3 transition.

Since such transitions result in either Auger electron emission or X-ray emission, X-ray notation is favored. For Auger processes, the resulting emission would be referred to as a KL_3L_2 electron, assuming the emitted electron came from the $2p_{1/2}$ level. For fluorescence, the emission would be referred to as a KL_3 X-ray, or, if the Siegbahn X-ray convention were to be used, this would be referred to as a $K\alpha_1$ X-ray. Siegbahn notation, named after Manne Siegbahn (the father of Kia Siegbahn), is most commonly used in the X-ray community. In this notation, the Greek letter denotes allowed transitions from stationary states of the next principal quantum number (the order being α, β, ζ), while the subscribed refers to the relative intensity (1 being the most intense).

A more complex example lies in the charge transfer that occurs on 2p core hole formation in the Cu^{2+} ion when bound within CuO. This process, described in Section 5.1.1.3.2.1, takes the form of electron transfer from the 2p level of an attached O^{2-} ion to the Cu-3d level. (Note: Such configurations are only noted during core hole formation.) In full spectroscopic notation, this would be represented as

$$\text{For } Cu^{2+}: 1s^2 2s^2 2p^5 3s^2 3p^6 3d^9 \rightarrow 1s^2 2s^2 2p^5 3s^2 3p^6 3d^{10}$$

$$\text{For } O^{2-}: 1s^2 2s^2 2p^2 \rightarrow 1s^2 2s^2 2p^1$$

To simplify matters, this can be represented for the ion of interest using only the levels affected. For the Cu^{2+} ion, this takes the form

$$2p^5 3d^9 \text{ L} \rightarrow 2p^5 3d^{10} \text{ L}^{-1}.$$

In this case, the ligand is represented by the term L (not to be confused with the term L used in X-ray notation for defining a specific stationary state, or the term l used to define the angular momentum) and the ligand minus one of its electron by the L^{-1} term.

This can be further simplified by using the c and c^{-1} terms to describe the core level to be affected and the core hole, respectively, and by dropping the numerals defining the principal quantum number of the stationary state experiencing charge transfer. For Cu^{2+}, this would take the form

$$c^{-1} d^9 \text{ L} \rightarrow c^{-1} d^{10} \text{ L}^{-1}.$$

This is the form used in Section 5.1.1.3.2.1. Note: Spectroscopic notation tends to be preferred in such cases since this more effectively relays the transitions occurring.

2.1.2.4 *Stationary States*

Electrons that are bound to an atom/ion only exist in specific, discrete energy levels referred to as stationary states. In ground-state atoms, stationary states closest to the nuclei are filled first. This is referred to as the *Aufbau principle*. Since the energies of the valence stationary states can overlap in the heavier elements, partially filled levels containing electrons with lower principal quantum numbers (n) can occur without violating this principle.

The term stationary state is used since it describes a level within a particular atom/ion whose energy does not vary with time (assuming no excitation/relaxation occurs). This observation is one of the cornerstones of modern quantum mechanics, also referred to as quantum wave mechanics. The classical view stipulates that a loss of energy should occur with time due to the radiation of energy (centrifugal force). If this were to occur, all electrons would spiral into their respective nuclei with their stationary states displaying a continuum of energies.

Core-level stationary-state energies (core-level *B.E.* values) do, however, vary for atoms with differing numbers of protons. The number of protons in an atom/ion defines the atomic number (Z) of the respective atom/ion. Z defines the element (Z is equal to the number of protons), as well as the number of electrons attached to a neutral atom (the number of protons and electrons must be equal to each other for electrical neutrality to prevail). Ions are atoms that have gained or lost one or more electrons.

The dependence of core-level *B.E.*s of a particular stationary state on Z for ground-state atoms can simplistically be understood as arising from the coulombic interaction that occurs between electrons and protons in close proximity to each other, and the fact that atoms with higher Z impart a greater attraction on their core electrons. In other words, the fixed number of core electrons (those in a specific core stationary state) experiences the attractive power of a greater number of protons as Z increases. This also reduces the separation between the respective core electrons and the nuclei (r). The *B.E.* dependence on Z scales roughly with $1/r^2$. This dependence is illustrated in Figure 2.3. The decrease in *B.E.* with increasing n stems from an increase in r, which can partially be attributed to the fact that the inner electrons shield the outer electrons from the attractive power imparted by the protons within their respective nuclei.

Core-level *B.E.* values are also influenced by the distribution of electrons within the various stationary states. This is a function of the state the atom exists in (electrons can exist in a ground state or in an excited state with the latter requiring energy), as well as the environment the atom/ion is in (bonding alters the energy of states, albeit by a small but measurable amount). As an example, adding or removing an electron in an atom has the effect of distributing the attractive forces of the protons within the respective nuclei over a greater or lesser number of electrons (the number of protons remains fixed). This explains why ions generally exhibit different *B.E.* values than their atomic counterparts. And lastly, variations are introduced as the values of the quantum numbers l, m_l, and m_s of the respective electrons are

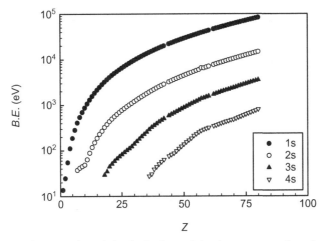

Figure 2.3. Binding energies of the 1s, 2s, 3s, and 4s electrons as a function of Z from all the naturally occurring elements from Li–Hg as measured by XPS and XRF (these values are listed in Appendix B).

altered. This stems from the introduction of nonsymmetric distributions and/or spin orbit splitting (the latter is discussed in Section 2.1.2.5).

This correlation between Z and $B.E.$ was first described theoretically by Arnold Sommerfeld in 1916. Although effective in approximating the $B.E.$s, the old quantum mechanics used (the same form employed in the Bohr model) required electrons to follow elliptical orbitals. The more correct quantum *wave* mechanics with relativistic effects was applied by Paul Dirac in 1928 to describe these $B.E.$s. Within this concept, electron distributions were modeled using density probability functions (the same as described by the Schrodinger equation). These functions imply that electrons do not follow well-defined orbitals; rather, they have a probability of being within a specific region in time. Note: Dirac's interpretation reduces to the Schrodinger equation at nonrelativistic velocities. Further discussion on these aspects can be found in Appendix C and in greater detail in various physics texts (e.g., see Eisberg and Resnick, 1985, Griffiths, 2004). Quantum wave mechanics is referred to as *quantum mechanics* throughout the remainder of this text.

2.1.2.5 Spin Orbit Splitting The fine structure noted in various X-ray and electron spectroscopies, inclusive of the XPS, AES, and XRF of bound electrons with the same n but nonzero l quantum numbers, arises from an effect termed spin orbit splitting (also called *spin orbit effect* or *spin orbit coupling*). An example of this is illustrated in

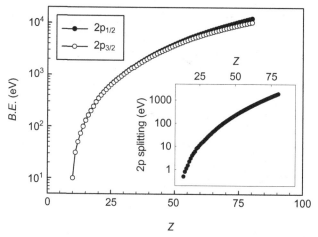

Figure 2.4. Binding energies of the $2p_{1/2}$ and $2p_{3/2}$ electrons as a function of Z for the naturally occurring elements from Li–Hg as measured by XPS and XRF (these values are listed in Appendix B). The difference between these levels, that is, the spin orbit splitting energy versus Z, is shown in the inset.

Figure 2.4 and its inset, in which the energy of the electrons with the $n = 2$, $l = 1$, and $m_s = 1/2$ is compared to those with $n = 2$, $l = 1$, and $m_s = -1/2$ (those that reside in the $2p_{1/2}$ vs. $2p_{3/2}$ stationary states). This is referred to as fine structure since the splitting in energy is minimal when compared to the $B.E.$ dependence on Z.

Spin orbit splitting stems from the coupling of the magnetic fields set up by an electron spinning around its own axis with that of an electron spinning around its nucleus if following a nonsymmetric orbit. Note: Every spinning charged body induces a magnetic field. Since the spin of an electron is defined by m_s and the symmetry of the orbital is expressed by l, two possibilities exist as expressed by j (recall: $j = l + m_s$). This is sufficient to influence the $B.E.$ of electrons from $l > 0$ levels.

The degeneracy (number of electrons in these two states) can be approximated using Russell–Saunders (L–S) or j–j coupling arguments (Condon and Shortley, 1935). The aforementioned applies to low Z elements (generally $Z < 30$), while the latter applies to high Z elements. This can be understood since in light elements, s (m_s) and l interact among themselves, that is, they can be treated separately, whereas in heavier elements, these couple. When coupling occurs (referred to as j–j coupling), the quantum numbers S and L lose their relevance. In any case, both arguments reveal that the number of electrons in these levels scales as $2J + 1$. As an example, twice as many electrons exist in the $n = 2$, $l = 1$, and $m_s = 1/2$ state as those in the $n = 2$, $l = 1$, and $m_s = -1/2$ state. The level with the larger j (same n and l) has the lower *B.E.*

The increased spin orbit splitting noted for electrons with the same n and l from elements with increasing Z stems from the increased interaction with the magnetic field induced by the nucleus. This splitting dependence on Z is in the inset of Figure 2.4.

This can be envisaged if the nucleus spins around the electron. Although this does not actually occur (lighter bodies always spin around heavier bodies), the effect will be the same as if the electron were spinning around the nucleus. A spinning nucleus would then set up a magnetic field proportional to the charge of the nucleus (dependent on Z) and distance to the electron.

2.2 SUMMARY

All matter, whether present in the gas, liquid, or solid states, comprises a grouping of atoms and/or ions. The properties exhibited by these groupings are dictated by the type of atoms and/or ions present and by the form in which they are bound. Bonding occurs via the interaction of the outermost electrons (valence electrons) from each atom/ion. XPS derives information on the elemental composition (what atoms are present) and speciation (how the atoms/ions are bound) by assessing the electronic structure of the atoms and/or ions residing within the surface region of the sample being analyzed.

The basis of XPS can be directly related to the predictions implied by quantum mechanics, with the spectra acquired revealing peaks of discrete energy consistent with those implied by quantum mechanics. Quantum mechanics is a mathematical interpretation that assigns each bound electron a specific set of quantum numbers when residing in a specific stationary state (element-dependent energy levels). The total number of electrons in a neutral atom (those among the various stationary states) also equates to the number of protons, with the latter defining the atomic number (Z) and hence the element. Ions are atoms that have gained or lost one or more electrons.

Since stationary states can only accept a finite number of electrons, every element has a specific electronic structure with the energy of the electrons in each stationary state having element-specific values (these increase roughly as $1/r^2$ with increasing Z, where r is the separation between the respective electrons and protons). Stationary states are defined via either spectroscopic or X-ray notation (both define the same thing). Transitions between these states can also be described via either notation or some derivative thereof. Additional *B.E.* variations occur with the electronic structure (due to shielding, excitation, and spin orbit splitting) and the local chemical environment in which the respective element exists (bonding).

The fact that XPS samples the core electronic structure of the atom/ion of interest (core electron *B.E.*s are derived) allows it to identify the elemental composition of any solid (all elements from Li to U are detectable if above 0.05 atomic % and within a depth of 10 nm from the surface), as well as the local chemical environment, that is, speciation of the aforementioned atom/ion.

CHAPTER 3

XPS INSTRUMENTATION

3.1 PREREQUISITES OF X-RAY PHOTOELECTRON SPECTROSCOPY (XPS)

Two characteristics define the effectiveness of most any advanced analytical technique, these being

(a) The instrumentation (this defines what is ultimately possible)
(b) The skill and knowledge of the person involved in collecting and deciphering the spectra produced

This chapter provides an introduction to the instrumentation in current use.

The ability to collect and decipher the spectra is covered in Chapters 4 and 5.

As illustrated in Figure 1.2, an XPS instrument primarily consists of

(a) An X-ray source
(b) Extraction optics and energy filter
(c) A detection system

X-ray Photoelectron Spectroscopy: An Introduction to Principles and Practices, First Edition. Paul van der Heide.
© 2012 John Wiley & Sons, Inc. Published 2012 by John Wiley & Sons, Inc.

Within this instrument, the environment must be such that

(a) The photoelectron and Auger electron emissions are not affected by any external electrostatic or magnetic fields.

(b) The photoelectron and Auger electron emissions are able to traverse the region between the sample and the detector (around 1 m in distance).

(c) The sample from which the emission arises must not be modified in any form or fashion during the course of analysis.

All external magnetic or electric fields must be minimized or accounted for; otherwise, they will affect the $K.E._{XPS}$ and hence the $B.E._{XPS}$ of the emitted electrons ($K.E._{XPS}$ and $B.E._{XPS}$ are defined in Section 1.4). X-rays are not influenced since these have no charge. External fields include, among other things, that arising from the earth's rotation, that is, the earth's magnetic field. Although this remains constant, it does vary from location to location around the globe.

For this reason, vendor installed instruments are set up to account for this field, which may require minor adjustments over time using routines discussed in Section 4.1.3. XPS instruments also use *mu-metal* (~76% Ni, ~17% Fe, ~5% Cu, and ~2% Cr) either in the construction of the analysis chamber, energy filter housing, and detector housing, or in the lining of these chambers/housings since this material acts as an effective shield against any extraneous magnetic or electrostatic fields. If used as a lining, 316L grade stainless steel (~68% Fe, ~17% Cr, ~12% Ni, and 3% Mo) is then used in the chamber/housings.

These materials are also used since they provide the conditions necessary for the production of ultrahigh vacuum (UHV) conditions (reasons for UHV are discussed further in Section 3.1.1). Note: Vacuum conditions are required in XPS to satisfy the remaining two criteria listed above. This can be understood since under atmospheric conditions, there exists on the order of 2×10^{19} molecules/cm^3; that is, such a density will, among other things, prohibit the transmission of electrons from the sample surface to the detector.

3.1.1 Vacuum

The word "vacuum" comes form the Latin word "vacuo," which means empty (vacuum and vacuo are to this day often interchanged). We now know, however, that there is no such thing as a completely empty space (even the vacuum of outer space contains a few atoms per cubic meter). A revised definition for vacuum can thus be considered as a region in space containing less gas than its surrounding regions.

TABLE 3.1   Some Useful Conversion Factors for Pressure

1 Torr (1 mmHg)	=133 Pa	=0.0193 psi	=1.30 mbar	=0.00132 atm

A vacuum is defined by the pressure, or lack of, within the region of interest. The SI unit for pressure is the *pascal* (Pa). That most commonly used unit in vacuum science and technology is the *Torr*. The *millibar* (mbar) is also heavily used. Conversion factors between these and other units of pressure are listed in Table 3.1. Another term of interest is the *langmuir* (L). This is the pressure–time integral required to form a monolayer; that is, 1 L equates to ~10^{-6} Torr·s assuming a surface with a unit sticking coefficient.

The importance of vacuum in XPS cannot be overstated. This is probably best illustrated when collecting spectra from a metal surface. Even on precleaned surfaces, a sizable C-1s peak is still noted. Quantifying this signal reveals levels ranging from 1 to 15 atomic %, depending on the sample and the sampled volume. This signal arises from the adsorption of gas-phase molecules that occurs even under UHV conditions.

To understand how this is possible, one first needs to realize that in any gas at room temperature and atmospheric pressure (also defined as 1 atm, 101.3 kPa, 760 Torr, etc.), there exists ~2×10^{19} molecules flying around in random directions at velocities that are in excess of 100 m/s (this is mass and temperature dependent) with the average velocity best described using the *kinetic theory of gases*. Since these molecules move, there is a finite probability that they will collide with some solid surface, that is, container walls and sample. Indeed, there are over 10^{23} collisions/cm²·s under atmospheric conditions, which is what results in the pressure measured.

The pressure (P) resulting from these collisions (as described via the kinetic theory of gases) is defined as

$$P = n \cdot k_B \cdot T, \tag{3.1a}$$

where n is the number of particles per unit volume (typically expressed in cubic centimeter), k_B is the Boltzmann constant (joule per Kelvin), and T is the temperature (Kelvin). From this relation, the velocity of the particles can be defined as

$$v = (8k_B \cdot T/\pi \cdot m)^{1/2}, \tag{3.1b}$$

where m is mass in unified atomic mass units. The dependence of pressure on temperature results from the increased energy (velocity) contained within the randomly traveling molecules. Upon substitution, the collision rate can be written as

$$Z_a = P/(2\pi \cdot m \cdot k_B \cdot T)^{1/2}. \qquad (3.2a)$$

This is the *Hertz–Knudsen* relation expressed in SI units. This can be expressed in units of Torr for P as

$$Z_a = 3.51 \times 10^{22} (P/(T \cdot m)^{1/2}). \qquad (3.2b)$$

This provides Z_a in units of collisions per square centimeter-second and reveals that at atmospheric pressure and temperature, each surface atom is struck in excess of 1×10^8 times per second. Applying this to a silicon surface (sticking coefficient close to unity and a surface density of 1×10^{15} atoms/cm^2) reveals that a contaminant monolayer forms within a few nanoseconds at atmospheric pressure and temperature. This drops to ~1 s at 10^{-6} Torr but does not really reach an acceptable value (~1000 s) until a vacuum of ~10^{-9} Torr is reached.

Another useful parameter is the mean free path of the molecule in the gas phase. As would be expected, this is dependent on the pressure, temperature, and radii of particles. The collisional cross section, σ, can be described by a cylinder of radius, r, such that all particles that fall within an area of πr^2 collide with this particle. Since both are traveling with respect to each other, a $\sqrt{2}$ factor is also inserted. In units of centimeter, the average distance traveled between collisions (the mean free path) is

$$d_{\text{M.F.P}} = 7.50 \times 10^3 (k_B \cdot T/(P \cdot \sqrt{2} \cdot \pi \cdot r^2)), \qquad (3.3)$$

where k_B is in units of joule per kelvin, P is in units of Torr, and r is in units of centimeter. Values of m, r (derived from is Lennard–Jones potential), σ, and the pressure at which the mean free path equals 1 m are listed, along with atomic mass, in Table 3.2.

The dependence of the collision rate along with the average distance traveled for O_2 molecules and ions as a function of pressure is summarized graphically in Figure 3.1. The dotted line represents a minimum acceptable vacuum. The mean free path for electrons equates to ~$(4 \times \sqrt{2})$ times that which would be derived via Equation 3.3.

Also shown in Figure 3.1 are the definitions used to describe the various pressure ranges (vacuum ranges). Since these are not universally accepted, those from the American Vacuum Society (AVS) are used. These are defined in Table 3.3.

Other aspects of note include

(a) Detection probabilities in XPS increase with improved vacuum (due to increased transmission resulting from fewer collisions)

TABLE 3.2   List of Molecular Mass (u), Collisional Cross-Section Radius (cm), Collisional Cross Section (cm²), and the Pressure (Torr) at Which the Mean Free Path Equals 1 m for the Listed Molecules (from Redhead et al., 1993).

Gas Molecules	m (u)	r ($\times 10^{-8}$ cm)	σ ($\times 10^{-15}$ cm²)	P ($\times 10^{-5}$ Torr)
H_2–H_2	2	2.556	2.05	8.03
N_2–N_2	28	3.681	4.26	5.07
O_2–O_2	32	3.433	3.70	5.84
Ar–Ar	40	3.418	3.67	5.88

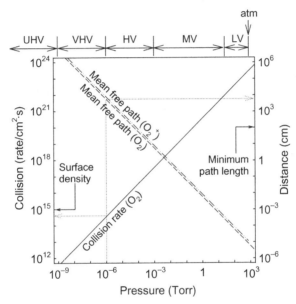

Figure 3.1. Surface collision rates (left ordinate) and mean free path (right ordinate) as a function of pressure for O_2 molecules and ions. The regions specified as LV, MV, HV, and UHV refer to *low vacuum, medium vacuum, high vacuum,* and *ultrahigh vacuum,* respectively. *Note*: Since these ranges are not universally agreed upon, the AVS definitions are used.

TABLE 3.3   Pressure Range Definitions as Used by AVS in Units of Pascal (the SI Unit for Pressure) and Torr (See Table 3.1 for Conversion Factors)

Nomenclature	Pressure (Pa)	Pressure (Torr)
Low vacuum (LV)	$1 \times 10^5 \leftrightarrow 3.3 \times 10^3$	$760 \leftrightarrow 25$
Medium vacuum (MV)	$3 \times 10^3 \leftrightarrow 1 \times 10^{-1}$	$25 \leftrightarrow \sim 1 \times 10^{-3}$
High vacuum (HV)	$1 \times 10^{-1} \leftrightarrow 1 \times 10^{-4}$	$\sim 1 \times 10^{-3} \leftrightarrow 1 \times 10^{-6}$
Very high vacuum (VHV)	$1 \times 10^{-4} \leftrightarrow 1 \times 10^{-7}$	$\sim 1 \times 10^{-6} \leftrightarrow 1 \times 10^{-9}$
Ultrahigh vacuum (UHV)	$1 \times 10^{-7} \leftrightarrow 1 \times 10^{-10}$	$\sim 1 \times 10^{-9} \leftrightarrow 1 \times 10^{-12}$
Extremely high vacuum (XHV)	$< 1 \times 10^{-10}$	$< \sim 1 \times 10^{-12}$

until a pressure of ~1×10^{-6} Torr is reached. Thereafter, little variation is noted.

(b) Adsorption rates on the sample surface being examined are enhanced as a result of the photon and/or electron impact. This is sample dependent and arises from the energy deposited over the region irradiated.

3.1.1.1 *Vacuum Systems*

Although the vacuum pump was invented in 1650, it took close to 300 years to produce vacuum conditions better than ~10^{-7} Torr. Indeed, this is one of the reasons why effective XPS instruments did not appear until the 1960s.

A vacuum is produced by reducing the amount of molecules within a confined space relative to its surroundings. Indeed, it is the presence of these molecules or, more precisely, their collisions with their surroundings that produce pressure. To reach UHV conditions, however, requires a highly specialized pumping system and a chamber displaying exceedingly low outgassing characteristics. Outgassing describes the removal of surface adsorbates generated when exposing any surface to atmospheric gases.

Vacuum chambers are typically constructed out of stainless steel with mu-metal lining or mu-metal itself. These materials are used due to their

(a) Low outgassing rate (once surface adsorbates have been removed)
(b) Low corrosion rate (this is further minimized under UHV)
(c) Low vapor pressure (stainless steel has a high melting point)
(d) Structural integrity (under both ambient and thermal conditions)
(e) Cost-effectiveness and relative ease of fabrication (machining)

The use of brass, borosilicate glass, epoxy resins, adhesive tapes, rubber O-rings, and other high vapor pressure materials is minimized.

To further enhance the structural integrity, chambers capable of supporting UHV conditions tend to be manufactured in cylindrical or spherical shapes. Access is then provided through ports attached using specialized welding procedures. Other units are then connected using specifically designed flange/gasket combinations.

The most common flange/gasket type used in UHV systems is what is referred to as the *Conflat* system. This system, trademarked by Varian and Associates, comprises oxygen-free high-conductivity (OFHC) copper gaskets, which are pressed into flanges bearing a knife edge. When compressed (bolted together), the flanges press into a copper gasket generating a UHV tight seal. Such gaskets must, however, be replaced following each use. Other soft gasket materials that have been used in producing UHV conditions include Al, Au, and In. For HV applications, carbon-based O-rings are commonly used since these tend to be more cost affective; that is, these are reusable.

Introduction of samples from the atmosphere is carried out through an attached self-contained introduction chamber that can be sealed off via a series of valves from the atmosphere and/or the analysis chamber. This allows the sample to be pumped down from atmosphere to a pressure of $\sim10^{-7}$ Torr within 10–20 minutes depending on the outgassing rates of the sample. Once pumped down, the valve between the introduction chamber and the analysis chamber is opened to allow sample transfer.

The plumbing used in XPS instruments comprises multiple pumps, valves, tubing (plumbing), and, of course, vacuum gauges for registering the vacuum produced. A typical example of a vacuum system is displayed in Figure 3.2. This reveals the use of several different types of vacuum pumps that, together with the valves and plumbing, are operated in specific sequences.

Vacuum pumps can be broadly subdivided into three groups, these being

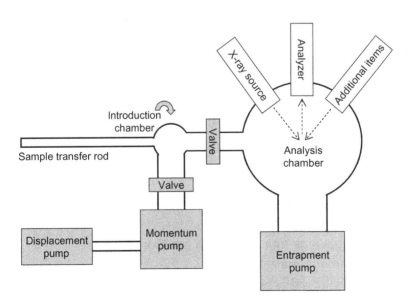

Figure 3.2. Schematic example of the vacuum plumbing used in a typical XPS system.

(a) Positive displacement pumps
(b) Momentum transfer pumps
(c) Entrapment pumps

Positive displacement pumps are best for evacuating from atmospheric pressure to ~10^{-3} Torr and for supporting (backing) momentum transfer pumps. Those most commonly found on XPS instruments are *rotary vane pumps*.

Momentum transfer pumps can produce a vacuum down to 10^{-10} Torr for chambers already pre-evacuated to within the 10^{-3} Torr range. The most commonly used momentum transfer pumps used in XPS instruments are *turbomolecular pumps*. These are typically set up, along with a positive displacement pump, to evacuate an introduction chamber. Note: Since an introduction chamber is routinely exposed to the atmosphere, the best pressure that can be reached within an acceptable time is in the 10^{-7} Torr range. This, however, is sufficient to allow a sample to be transferred into the analysis chamber held under UHV (once the valve between the two is opened), that is, it does not allow significant contamination of the analysis chamber.

Entrapment pumps are self-contained units (these do not require any other pumps) capable of producing a vacuum down to 10^{-10} Torr. As the name suggests, entrapment pumps operate by trapping or condensing gas-phase molecules into the solid state or by ionization and accelerating these molecules into a solid. Those most commonly found in state-of-the art XPS instruments are *ion pumps* with *titanium sublimation* pumps often added. Ion pumps are operated continuously to sustain UHV conditions. Sublimation pumps, on the other hand, are operated in an intermittent fashion to allow improvement in vacuum conditions. Since neither should not be operated at pressures greater than ~10^{-5} Torr, the initial pumpdown of these chambers is carried out through a combination of displacement and momentum transfer pumps.

Pumping speed must also be considered when designing/constructing a chamber capable of supporting UHV conditions. The primary factors controlling pumping speed are

(a) The chamber volume (smaller chambers provide faster pumpdown rates)
(b) The outgassing rate, which is a function of
   (i)   The materials used
   (ii) The adsorbates present
   (iii) The surface area
(c) The ultimate pressure desired (10^{-9} Torr or better)

Pumpdown times from VHV to UHV are by far the longest. This arises since in this regime, it is the removal rate of surface adsorbates that controls the pressure. Adsorbates develop when any metal surface is exposed to atmospheric gases. Indeed, under ambient conditions, it can take weeks to pump a vacuum chamber from the atmosphere to UHV irrespective of the pumping system used. Note: This is also the reason samples are not transferred directly from atmosphere into analysis chambers.

To accelerate the removal of surface adsorbates, such chambers are baked whenever exposed to the atmosphere. This entails heating the chamber to 150–200°C for periods of between 12 and 48 hours while being pumped. For this, all temperature-sensitive units and connections must be disconnected.

The effect of outgassing can be represented as

$$V(dP/dt) = G_r - P \cdot S_p, \tag{3.4a}$$

where V is the volume of the chamber in units of liter, G_r is outgassing rate in units of liter-Torr per second, P is pressure in units of Torr, and S_p is the pumping speed in units of liter per second. A good rule of thumb for clean stainless steel chambers is to provide at least 1 L/s pumping speed per 100 cm^2 of surface area. Since typical XPS instruments have a surface area of $\sim 2 \times 10^4$ cm^2, the pumping speed of the entrapment pump is typically around 200 L/s.

The ultimate pressure that can be reached can be derived from Equation 3.4a. This takes the form

$$P = G_r/S_p. \tag{3.4b}$$

Lastly, the ability to measure the vacuum derived must also exist. In XPS, this is most commonly carried out using a combination of two types of gauges, these being

(a) The Pirani gauge (useful pressure range is ~ 10 to $\sim 10^{-3}$ Torr)
(b) The ionization gauge (useful pressure range is $\sim 10^{-3}$ to $\sim 10^{-10}$ Torr)

These are located within XPS instruments such that any emissions that may be produced from these detectors cannot influence the recorded spectra. This is accomplished by placing them "out of the line of sight" of the sample surface and the detector.

3.1.2 X-ray Sources

Since XPS is concerned with the analysis of core electrons from a solid surface, sources used in XPS must be able to produce photons of a sufficient energy to access a suitable number of core electron levels. Photons of this energy lie within the X-ray region of the electromagnetic spectrum. As a result, these are otherwise referred to as *X-rays*.

X-rays used in commercially available XPS instruments are produced via one of two sources, these being, in the order of usage,

(a) X-ray tubes with the different geometries used referred to as
    (i) Monochromatic sources
    (ii) Standard sources
(b) Synchrotron sources

Ultraviolet photoelectron spectroscopy (UPS) uses a third type of source referred to as a gas discharge lamp. These produce much lower-energy photons, that is, those lying in the ultraviolet (UV) region of the electromagnetic spectrum. Since UPS is often also carried out in commercially available XPS instruments, these sources are discussed in Section 3.1.2.3.

X-ray tubes produce X-rays by directing a sufficiently energetic electron beam at some metallic solid. This metallic object is referred to as the *X-ray anode*, with the electron source being the cathode. Cathodes comprise a thermionic source as discussed in Section 3.1.3.1. Although any solid can in principle be used as an X-ray anode, Al has become that most commonly used in XPS. This popularity stems from

(a) The relatively high energy and intensity of Al-$K\alpha$ X-rays
(b) The minimal energy spread of Al-$K\alpha$ X-rays
(c) The fact that Al is an effective heat conductor
(d) The ease of manufacture and use (robustness) of such anodes

The high intensity results in part because, in the case of Al anodes, $K\alpha$ X-rays are the most prominent of the X-ray emissions produced. Note: $K\alpha$ X-rays actually comprise $K\alpha_1$ and $K\alpha_2$ X-rays, but due to their small energy separation, these tend to be described together as either $K\alpha_{1,2}$ X-rays or more simply as $K\alpha$ X-rays. These arise from a multistep process as follows:

(1) An Al-1s electron is ejected as an incoming energetic electron scatters inelastically (Al-1s $B.E._{XPS}$ is 1559.6 eV).
(2) The Al-1s core hole produced is then filled by an electron from a 2p level (Al-$2p_{1/2}$ and $2p_{3/2}$ $B.E._{XPS}$ values are 73.0 and 72.5 eV, respectively).

(3) The 1s-2p energy difference, minus some perturbation energy difference (~0.4 eV for Al), is removed via Auger electrons ($KL_{2,3}L_{2,3}$ emissions) or fluorescence ($K\alpha_{1,2}$ X-ray emissions).

Al-$K\alpha_{3,4}$ and Al-$K\beta$ X-rays, Bremsstrahlung radiation (photons spanning some continuous energy range that arise from the deflection of electrons by surrounding charged particles), and other Auger electron emissions are also produced albeit to much lesser intensities.

The minimal energy spread is of importance since this is the primary factor that governs the energy resolution in XPS and, hence, the ability of XPS to relay speciation information. The energy resolution is a function of the X-ray source line width, the intrinsic width of the photoemission and the energy resolution of the energy analyzer. The source width, however, has by far the greatest influence as discussed further in Section 3.1.5.4. Current state-of-the-art instruments using monochromatic Al-$K\alpha$ sources along with concentric hemispherical analyzers (CHAs) are capable of an energy resolution of between 0.3 and 0.5 eV when operated in the *constant analyzer energy* (CAE)/*fixed analyzer transmission* (FAT) mode (see Section 3.1.5.3).

The requirement of being a good heat conductor is realized in the fact that, in order to maximize the X-ray flux, the energy of the electrons impinging on the X-ray anode should be set to ~$10 \times E_p$ where E_p is the X-ray (photon) energy. For Al-$K\alpha$ sources where E_p equates to 1486.6 eV, electron impact energies of 15 keV are thus used with the anode held at +15 kV and the remainder of the instrument grounded.

Such energetic electron impact, however, results not only in the production of X-rays but also of secondary electrons, Auger electrons, Bremsstrahlung radiation, and so on. Since the X-rays produced only account for a few percent of the energy deposited, these anodes will rapidly decompose if the heat produced is not effectively dissipated. To extend anode lifetimes, water cooling is also implemented. Due to the high voltages applied, high-purity water (deionized water) must, however, be used to avoid electrical discharge issues (arcing) between the anode and the remainder of the source.

A list of some anode materials along with the energy and energy spread of the respective X-ray lines produced is given in Table 3.4. Siegbahn notation is used since this is the most commonly applied by X-ray spectroscopists (see Section 2.1.2.2). Note: Only a few of these are used in XPS with the Al-$K\alpha$ X-rays being by far the most common.

3.1.2.1 *Standard Sources* These are fairly crude units that essentially consist of a water-cooled anode (attracts negative charge), which

TABLE 3.4 Some Anode Materials Used in Producing X-rays along with the Energy and Energy Spread (FWHM) of the Emissions Produced (Nonmonochromated)

Anode	Z	Emission	Energy (eV)	Energy Spread (eV)
Mg	12	Mg-$K\alpha$	1253.6	0.70
Al	13	Al-$K\alpha$	1486.6	0.85
Si	14	Si-$K\alpha$	1739.6	1.0
Ti	22	Ti-$K\alpha$	4510.9	2.0
Cr	24	Cr-$K\alpha$	5417.0	2.1
Ni	28	Ni-$L\alpha$	851.5	2.5
Cu	29	Cu-$K\alpha$	8048.0	2.6
Y	39	Y-$M\zeta$	132.3	0.47
Zr	40	Zr-$M\zeta$	151.4	0.77
Nb	41	Nb-$M\zeta$	171.4	1.21
Mo	42	Mo-$M\zeta$	192.3	1.53
Ag	47	Ag-$L\alpha$	2984.2	2.6

faces the target (sample). Irradiation of this anode with energetic electrons then results in an unfocused X-ray beam that floods the entire sample and any surrounding areas in close proximity. Dual Al and Mg anodes are commonly present in these sources.

The primary attributes of standard sources when compared to monochromatic sources include

(a) Their low cost
(b) The freedom to use any anode material
(c) The ease of switching between anode materials (two anodes are typically set up on the X-ray source at any one time with these typically being Al and Mg)

When compared to monochromatic sources, disadvantages include

(a) A larger energy spread (when using Al and Mg anodes, both the $K\alpha_1$ and $K\alpha_2$ lines overlap, thereby producing a single peak of larger full width at half maxima [FWHM])
(b) The presence of additional peaks ($K\alpha_{3,4}$ and $K\beta$ as well as Bremsstrahlung radiation). Although minor in comparison to the $K\alpha_1$ and $K\alpha_2$ emissions, the latter can produce additional peaks (referred to as *ghost peaks*) in associated XPS spectra. These appear at slightly lower $B.E._{XPS}$ values than the respective main peaks. This results from the fact that $B.E._{XPS}$ values are

calculated via Section 2.1 for the dominant X-ray signal, that is, $K\alpha_1$ for Al and Mg.

(c) Greater sample degradation can result. This arises from Bremsstrahlung radiation and electrons allowed to strike the sample if the X-ray transparent foil used to quench these emissions is compromised (this comprises a ~2-μm thin Al foil). In addition, thermal exposure will result when the source is brought too close to the sample (see next point).

(d) Their lower photon flux per unit area (due to the unfocused nature of these sources). To compensate for this, the position of these sources relative to the sample is made adjustable such that these can be moved closer to the sample. If brought too close to the sample, this can restrict the access of other sources (electron and ions beams) to the sample.

Note: The lower flux per unit area of these sources can, in some cases, prove advantageous when dealing with insulating samples. This is discussed in Section 4.1.3.

3.1.2.2 Monochromated Sources These sources consist of the elements found in a standard source (water-cooled anode at which energetic electrons are directed) but with the X-rays reflected off a concave single crystal as illustrated in Figure 3.3.

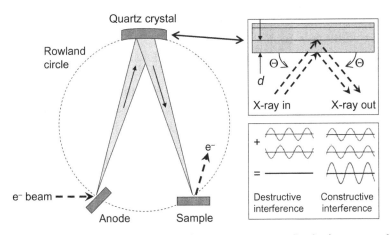

Figure 3.3. Schematic of a monochromatic source geometry. In the insets are shown the diffraction criteria; that is, only those X-rays with specific energy (wavelength) will reflect off a crystal at some specific angle defined by the crystal lattice spacing (Bragg relation). *Note*: Reflection only occurs when constructive interference between two waves arises.

Monochromatic sources found on commercially available XPS instruments use quartz single crystals since

(a) These are relatively inert under atmospheric conditions (extensive surface modification or buildup of surface adsorbates is not noted) as well as UHV compatible; that is, they do not outgas or degrade.

(b) An integer number of Al-$K\alpha_1$, Ag-$L\alpha_1$, Ti-$K\alpha_1$, and Cr-$K\beta_1$ X-ray wavelengths (0.833, 0.415, 0.312, and 0.208 nm, respectively) equates to the spacing between the quartz I0$\bar{1}$0 planes (0.425 nm) at an angle close to 78.5°.

The latter allows the Bragg diffraction criteria (described below) to be satisfied to first-order for Al-$K\alpha_1$, second order for Ag-$L\alpha_1$, third order for Ti-$K\alpha_1$, and fourth order for Cr-$K\beta_1$ X-rays. Note: A significant decrease in intensity is noted with increasing order. As a result, Ti-$K\alpha_1$ and Cr-$K\beta_1$ emissions are rarely used in XPS.

The Bragg diffraction criteria (Bragg, 1913) stipulates that photons can only be reflected constructively if

$$n\lambda = 2d\sin\varphi, \tag{3.5}$$

where n is the diffraction order, λ is the wavelength, d is the crystal atomic spacing, and φ is the angle of diffraction. Since λ, and hence the energy, of the Al-$K\alpha_1$ and $K\alpha_2$ X-rays is different, the latter is filtered out. This, combined with a slight concavity introduced into the quartz crystal, results in a focused X-ray beam at the sample surface with a narrower energy spread than possible with a standard source (between 0.2- and 0.3-eV FWHM for Al-$K\alpha_1$ and ~1.2-eV FWHM for Ag-$L\alpha_1$). The radius of curvature is defined by the distance from the source and sample to the crystal face by what is referred to as the *Rowland sphere*. X-ray spot sizes from such sources range from ~10 μm to several thousand micrometers in diameter.

The main advantage of these sources when compared to standard sources lies in

(a) Their ability to produce a focused X-ray beam without the requirement of the source being in close proximity to the sample
(b) The narrower energy spread of the resulting X-rays

The primary disadvantages when compared to standard sources include

(a) Their higher cost
(b) The limited number of anodes allowable within a specific geometry needed to satisfy the Bragg criteria
(c) The greater potential for sample charging when analyzing insulating surfaces (a result of the greater photon flux per unit area relative to standard sources)

3.1.2.3 *Gas Discharge Lamps* Although these are not useful for producing core-level photoelectron emissions, these sources are discussed since

(a) These provide valence band spectra to significantly greater sensitivity and improved energy resolution than achievable using either the standard or monochromated sources.
(b) These are often found on nonsynchrotron-based XPS instruments equipped with either or both standard and monochromated X-ray sources.

The improved sensitivity arises from the closer match of the X-ray energy with the valence band energy. The improved energy resolution arises from the narrower energy spread of the photons produced (FWHMs are of the order of 0.005 eV). These cannot, however, be focused to the spot sizes attainable in monochromated sources.

Gas discharge lamps used in UPS pass He gas through the discharge region, which is formed by applying a potential difference between the anode and cathode regions of ~1 kV (the cathode is earthed). This causes the excitation of He electrons to higher stationary states. On de-excitation, photons of energy equal to the energy difference between the states accessed are released. The primary emission at 21.2 eV is that most commonly used in UPS. A secondary, weaker emission at 40.8 eV is also accessible but with effort required (cleanliness of chamber and gas handling system is imperative).

3.1.2.4 *Synchrotron Sources* These sources produce photons by forcing electrons traveling at near relativistic speeds into curved paths (this is carried out using magnetic fields). The photons are emitted tangential to the electron beam with an intensity proportional to the electron beams' radius of curvature and inversely proportional to the cube of the electron's energy. The energy of the photons is continuous over ~1 eV to ~10 keV.

Such sources have the advantage of

(a) Allowing any photon energy to be selected (this provides, among other things, resonance effects associated with ionization thresholds to be examined)
(b) Providing a beam of narrower energy spread (this allows for improved overall energy resolution with a value of ~0.1 eV or better)
(c) Yielding a much higher photon flux per unit area than possible in standard or monochromated sources (a factor of ~100× or more)
(d) Producing a highly collimated beam that can be focused to ~0.1 μm through the use of zone plates (this allows for the ultimate in spatial resolution)
(e) Allowing a polarizable beam to interact with the sample (the polarization vector is along the plane of the circulating electrons)

The primary disadvantage of synchrotron sources lies in the fact that

(a) A synchrotron is needed (although available at various research institutions, their time is limited and can be costly).
(b) Greater sample degradation can occur during analysis particularly to organic samples (a result of the greater flux per unit these sources produce)

The higher flux per unit area, preselectable energy range, and superior energy resolution when studying core-level emissions do, however, make these the sources of choice in research facilities examining the frontiers of condensed matter physics, etc.

3.1.3 Electron Sources

Electron sources are used in XPS as the cathode within X-ray tubes (both standard and monochromated X-ray sources) and as an independent unit for directing electrons at insulating samples to aid in charge compensation (see Section 4.1.3). Although a range of electron sources exists, thermionic sources are those most commonly found in commercially available XPS instruments.

3.1.3.1 Thermionic Sources As the name suggests, thermionic sources generate electrons by heating a material to the point where the bound electrons have enough energy to overcome the material's work function (ϕ). When this occurs, they escape into the vacuum whereupon they can be directed to the sample at the desired energy (typically within the 10 eV–10 keV range in XPS) as dictated by the subsequent fields applied. Note: Due to space charge effects, beam diameters for all sources increase with decreasing beam energy.

The two primary thermionic sources in use are W filaments (ϕ is ~4.5 eV) and LaB_6 crystals (ϕ is ~2.6 eV). Due to the lower ϕ, LaB_6 sources provide higher currents while operated at lower temperatures (~2000 K for LaB_6 with respect to ~2700 K for W). This lower operating temperature results in a lower energy spread (~0.8 eV relative to 1.5 eV for W), which improves the lifetime. A lower energy spread also allows focusing to smaller beam diameters (spot sizes of <0.1 µm at 10 keV are possible with LaB_6 sources with respect to a few micrometers for W filaments). LaB_6 sources are, however, more sensitive to vacuum conditions (minimum operating pressures for W and LaB_6 sources are ~10^{-5} and 10^{-7} Torr, respectively) and are more costly.

3.1.4   Ion Sources

Ion sources are used in XPS to allow for sputtering of the surface of interest (removal of the surface layers as discussed in Section 4.3.1.2) or to aid in charge compensation when analyzing insulating samples (see Section 4.1.3). Although a range of ion sources exists, EI sources are those most commonly found in commercially available XPS instruments.

3.1.4.1   EI Sources   Sputtering in XPS has, to date, been most commonly carried out using inert gas ion electron impact (EI) sources. These sources consist of a region in which the gas of interest (Ar) is introduced and irradiated by electrons of ~70 eV from a heated filament. This induces the emission of electrons from the respective gas atoms/molecules, thereby producing the respective positive ions. The ions are then extracted through the orifice of a plate held at some negative potential relative to the ionization region.

To ensure the purity of this beam, a small bend along with a Wien filter are placed in the ion column. Wien filters consist of crossed electrostatic and magnetic fields that cause a deflection of the ion beam according to its mass, energy, and charge. This effectively removes the unwanted neutral components (atoms/molecules formed as a result of ion neutralization) as well as any multiply charged ions from the beam. Note: Both are undesirable since both have differing focusing characteristics. Electrostatic lenses and deflectors then direct a focused ion beam to the desired location on the sample. The deflectors also allow for rastering, that is, scanning of the beam over prespecified areas. These are set much larger than that analyzed by the X-ray beam to ensure an effective depth resolution can be attained.

Such sources provide medium brightness beams of ~10 eV to 20 keV and up to ~500 nA in current. The lifetime of these highly reliable sources is governed by the filament, which can be easily replaced. The only drawback, albeit minor in XPS, lies in the inability to focus the beam to spot sizes less than ~10 µm.

More recently, sputtering of organic layers via various cluster ion beams has attracted interest due to the significant reduction in sample damage observed relative to inert gas ion beams. The two cluster ion types of interest are organic cluster ions and inert gas cluster ions (Mahoney, 2009).

Organic cluster ions can be in the form of C_{60}^+, C_{84}^+, $C_{24}H_{12}^+$ (coronene), and so on. These use sublimation ovens to produce the required gas (coronene gas is produced when the powder is heated to ~170°C), which is then ionized via electron impact. These sources are capable of producing low-brightness beams of up to 20 keV (40 and 60 keV for doubly and triply charged ions) to currents approaching 50 nA (lower currents for doubly and triply charged ions) with the lifetime governed by the filament and the source material. The primary drawback of these sources lies in their increased cost relative to inert gas ion sources and their inability to focus a cluster ion beam to spot sizes less than several hundred micrometers.

Inert gas cluster ions comprise several thousand Ar atoms per singly charged ion. These are produced in a gas cluster ion beam (GCIB) source by condensing individual gas atoms into neutral clusters through cooling in a supersonic jet. These are then ionized via electron impact.

3.1.5 Energy Analyzers

Since information in XPS is derived from the $K.E._{XPS}$ of the electron emissions, effective speciation analysis requires an energy filter that exhibits both a high-energy resolution and a high transmission. The former allows for the separation of closely spaced peaks, thereby optimizing speciation identification capabilities, while the latter allows for sensitivity to be maximized. The two primary energy filter configurations used in XPS, illustrated in Figure 3.4a,b, include

(a) *Cylindrical Mirror Analyzer (CMA)*: These are common in AES instruments and older multitechnique instruments.

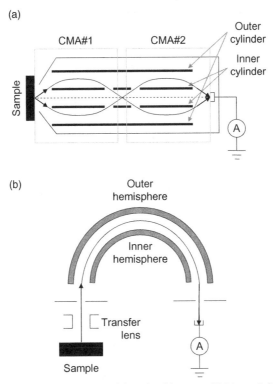

Figure 3.4. Schematic illustrations of (a) a double-pass CMA and (b) a CHA, along with the trajectories of electrons of the required energy within these units.

(b) *Concentric Hemispherical Analyzer (CHA)*: Also referred to as *hemispherical sector analyzer (HSA), spherical deflection analyzer (SDA)*, or *electrostatic hemispherical analyzer (EHA)*. These are by far the most commonly used in modern dedicated XPS instruments.

The primary advantage of a CHA relative to a CMA lies in the improved energy resolution possible; that is, values better than 0.01 eV are attainable. Of note is the fact that the energy resolution of a CHA is much better than the X-ray source energy spread. As a result, the energy resolution in XPS is primarily a function of the X-ray source, not the CHA. This is discussed further in Section 3.1.5.4.

A CHA also exhibits a weaker *K.E.* dependence on the distance between the analyzer and the sample (values of ~0.1 eV/mm common) than a CMA and can analyze larger areas than permissible when using a CMA. A CHA, however, has a poorer transmission due to the more limited collection solid cone than a CMA. This can be compensated for, to some extent, through the introduction of a transfer lens prior to the energy analyzer (not possible in a CMA) and the use of multichannel detectors (these simultaneously recorded the signal over multiple energy regions as discussed in Section 3.1.7.1). Transfer lenses also allow for greater accessibility to the sample's surface.

At this point, it is also worth mentioning a specific type of energy analyzer, termed the spherical mirror analyzer (SMA). This energy analyzer, used in parallel imaging (see Section 3.1.7.2), is similar to a CHA in that both utilize 180° hemispheres between which the electrons pass. The difference, however, lies in the fact that the radii of the trajectories of the desired electrons are significantly smaller than the inner sphere radius of an SMA. Openings are thus placed in the inner sphere to allow access.

3.1.5.1 *CMA*   A CMA consists of two cylinders, one inside the other, whose axes intersect the area analyzed and the detector as shown in Figure 3.4a. Focusing to second order is attained by applying a potential difference to the two cylinders for electrons emanating at a takeoff angle of 47.7° with respect to the sample surface. Although this results in a higher transmission than a CHA, a poorer energy/angle resolution and a greater sensitivity of the $K.E.$ to sample position are noted. The energy resolution, which typically lies in the 1.0- to 0.1-eV range, is fixed by the dimensions of the slits appearing in the flight path (inner cylinder). The voltage, V, applied to the inner and outer cylinders, scale as (from Zashkvara et al., 1966)

$$V = (E_o \ln(R_{out}/R_{in}))K_{CMA}, \qquad (3.6)$$

where E_o is the energy of the electrons entering the cylinders, K_{CMA} is a geometric constant, and R_{out} and R_{in} are the radii of the inner and outer cylinders, respectively. R_{in} is fixed by the linear distance between the sample surface and the detector (L_o); that is, this equals 6.1 R_{in}. These come in the form of single- or double-pass CMAs. In a single-pass CMA, the electrons cross the cylinder axis once, whereas in a double-pass CMA, the electrons cross the cylinder axis twice.

3.1.5.2 *CHA*   The CHA consists of two concentric (180°) hemispheres, one inside the other, of radius R_{in} and R_{out}. The area being analyzed and the detector are situated such that the path of electrons departing the sample and arriving at the detector forms lines tangential to the average of R_{in} and R_{out} at either ends of the hemispheres. A schematic example is shown in Figure 3.4b.

Applying specific potentials to these hemispheres results in the deflection of electrons of some specific E_o onto the detector. These potentials scale as (from Purcell, 1938)

$$V_{in} = E_o[3 - 2(R_{el}/R_{in})] \qquad (3.7a)$$

$$V_{out} = E_o[3 - 2(R_{el}/R_{out})], \qquad (3.7b)$$

where R_{el} is the average radius equal to $(R_{out} + R_{in})/2$. The remaining parameters take on the same meaning as for the CMA with the potentials, and hence the energy resolution, dependent on the $K.E.$ of electrons entering the hemispheres (the beam axis is parallel to R_{el}).

3.1.5.3 *Modes of Operation*   An energy spectrum in XPS can be attained via one of two modes. These are by

(a) Scanning the inner and outer cylinder/hemisphere potentials of the respective CMA/CHA across the E_o range. This is typically referred to as the *constant retard ratio* (CRR) or the *fixed retard ratio* (FRR) mode. This is the only option available in single-pass CMAs.

(b) Accelerating/decelerating the electrons to some constant E_o value before entering the CMA/CHA. This is referred to as the *constant analyzer energy* (CAE) or *fixed analyzer transmission* (FAT) mode. Although this is most commonly applied in instruments using CHAs with the acceleration/deceleration carried out in transfer lens, this can also be applied in instruments using double-pass CMAs. In both cases, the potentials applied to the hemispheres of the CHA, or the second stage of a double-pass CMA, are held constant.

In the CRR/FRR mode, the energy resolution decreases with increasing E_o and hence $K.E.$ As a result, the energy resolution is expressed as a percentage of $K.E._{XPS}$. High-quality CMAs are capable of a resolution approaching 0.1%. This translates to 1 eV for 1000-eV electrons.

In the CAE/FAT mode, the energy resolution remains independent of $K.E._{XPS}$ (E_o is held at some constant value) and can thus be expressed in units of E_o. This is commonly referred to as the *pass energy*, with lower values resulting in improved resolution, but at the cost of sensitivity. High-energy resolution studies in current state-of-the art instruments using monochromatic Al-$K\alpha$ sources along with a CHA carried out at pass energies of ~5 eV are thus capable of an energy resolution down to ~0.4 eV. To obtain this resolution for 1000-eV electrons in the

CRR/FRR mode would require a CHA with a radius well in excess of 1 m.

3.1.5.4 *Energy Resolution*   The energy resolution in XPS (ΔE) is typically defined as that resulting from the convolution of three parameters, these being

(a) The FWHM of the source (ΔE_p)
(b) The line width of the photoelectron emission (ΔE_n)
(c) The energy resolution of the energy analyzer (ΔE_a)

On the assumption that all components exhibit a Gaussian line shape, ΔE equates to

$$\Delta E = [(\Delta E_p)^2 + (\Delta E_n)^2 + (\Delta E_a)]^{1/2}. \tag{3.8}$$

ΔE_p values are dependent on the photon source used. Most commercially available XPS instruments are now supplied with monochromated Al-$K\alpha$ sources. These are capable of providing ΔE_p values of between 0.2 and 0.3 eV. Nonmonochromated Al-$K\alpha$ and Mg-$K\alpha$ sources (standard sources) have ΔE_p values equal to their intrinsic line widths, that is, those listed in Table 3.4 (0.85 and 0.70 eV, respectively). Only synchrotron and gas discharge sources can provide improved ΔE_p values, although the latter is only suitable for UPS.

ΔE_n values vary over a large range extending from <0.1 to >1.0 eV. These values, which are intrinsic to the sample being analyzed, arise from several factors including

(a) The irreducible line width (Γ), which reflects the core hole lifetime. This contribution is defined from the uncertainty principle (see Section 5.1.1)
(b) Line broadening introduced by phonons (lattice vibrations) along with other final state effects (these are discussed in Section 5.1.1.3)

The product results in a Gaussian–Lorentzian line shape with Γ introducing the Lorentzian component and the broadening introducing the Gaussian component (these different line shapes are discussed in Appendix D).

Lastly, ΔE_a values depend on the analyzer used (CHA or CMA as discussed in Sections 3.1.5.1 and 3.1.5.2, respectively) and the conditions under which these are operated (see Section 3.1.5.3). As an example, a high-quality CHA operated under the CAE/FAT mode with low-pass energy can yield ΔE_a values approaching 0.001 eV. A high-quality single-pass CMA, on the other hand, can yield ΔE_a values of ~0.1% (1 eV at a $K.E._{XPS}$ of 1000 eV).

In practice, ΔE is defined in XPS by measuring the FWHM of some core-level peak with a minimal ΔE_a. The Ag-3d peak satisfies this criterion and, as a result, is often used in deriving the specification of commercially available XPS instruments. Indeed, high-quality instruments provided with monochromated Al $K\alpha$ sources operated in the CAE/FAT mode at a pass energy of 5–10 eV provide ΔE values between 0.3 and 0.5 eV with values continually improving. Under optimal conditions (analysis of peaks with narrow FWHM on a well-aligned instrument), this can allow for the elucidation of $B.E._{XPS}$ values to within 0.1 eV.

Improved ΔE values can only be derived when using a gas discharge source or a synchrotron source. For example, UPS carried out at 20 eV using a high-quality CHA operated in the CAE/FAT mode at a minimal pass energy can yield a ΔE of ~0.003 eV. Note: This is approaching thermal scales; that is, it is equivalent to a temperature of 30 K. Likewise, XPS carried out at 500 eV using similar analyzers/conditions can provide a ΔE of ~0.05 eV (Hüfner, 2003).

3.1.6 Detectors

In XPS, it is not only important to measure the energy of the electron emissions but also the number of electrons produced. Indeed, XPS spectra are plotted in units of energy versus intensity, with the energy defined by the energy analyzer used (see Section 3.1.5) and the intensity defined by the number of electrons recorded by the detector. To obtain the best possible sensitivity, the detector must be capable of recording individual electrons, that is, operating in pulse counting mode. This signal is recorded in units of current (A), which are then represented in units of counts per second.

The current represents the passage of electrons (charge) across some fixed point in space per unit time (units of second). The charge carried by a single electron is -1.602×10^{-19} C. The passage of this single electron per unit second thus equates to 1.602×10^{-19} A. The sensitivity of conventional pulse counting electronics, however, lies in the 10^{-15} A range (this represents the passage of 6241 electrons per second). In order to record individual electrons, the detector used must thus exhibit a gain (multiplication factor of the original signal) of at least 10^4. The electron multiplier (EM) not only satisfies this criteria but can also be operated in the UHV conditions required by XPS.

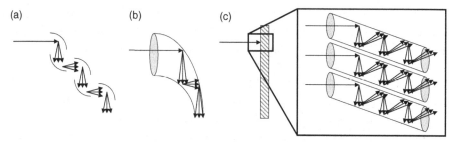

Figure 3.5. Pictorial representations (not to scale) of different EM geometries referred to as (a) discrete dynode detectors, (b) channeltron detectors, and (c) MCP detectors with three microchannels magnified (right). The arrows represent the incoming electrons along with the many fold (>10) electrons produced per collision (for the sake of clarity, not all are shown).

3.1.6.1 EMs

EM represents a group of detectors that operate by amplifying the signal generated by an electron impinging on the first or outermost surface of the EM, that is, that exposed to the incoming electron beam. The different EM configurations are shown in Figure 3.5. These include

(a) Discrete dynode EMs
(b) Channeltrons, also referred to as channel electron multipliers (CEM)
(c) Microchannel plate (MCP) detectors

A discrete dynode EM comprises a metallic conversion electrode followed by individual electrically isolated acceleration electrodes (also metallic). These electrodes are referred to as *dynodes*. The incoming electron strikes the conversion dynode that spans an area of ~10 × 10 mm and in doing so produces numerous secondary electrons. The secondary electrons are then accelerated into the next closest electrode (the first acceleration electrode) that produces more secondary electrons, which are then accelerated into the next acceleration dynode (second acceleration electrode), and so forth. These are capable of a gain of up to 10^9. These detectors, however, start to saturate when the incoming electron count rate exceeds ~1×10^6 cps. This results in a loss of input to output signal linearity.

A channeltron or CEM is a horn-shaped continuous dynode structure that is coated on the inside with an electron-emissive material such as PbO. An electron striking the opening of the channeltron thus creates an electron cascade in the same manner as in a discrete dynode EM. In this case, the electrons are accelerated into the horn by the potential difference that exists at both ends (2–4 kV). Although exhibiting many similar characteristics to discrete dynode EMs (similar linear range, etc.), channeltrons are capable of gains of up to 10^8. The prime advantage of these detectors with respect to XPS lies in the ability to shape the opening (by the manufacturer) such that an array of ~5 × 15 mm detectors, for example, can be placed side by side along a specific plane of interest, that is, the energy-dispersive axis of an energy analyzer. This 1-D array allows data to be acquired simultaneously over some energy range, which can then be plotted as a function of energy, or all of the data within some prespecified energy range can be added together to enhance overall sensitivity.

An MCP can be thought of as an extension of a 1-D channeltron array in that these consist of a 2-D array of parallel glass capillaries whose inner diameter ranges from 10 to 25 μm, and whose axis lies ~7° off that of the trajectory of the signal to be recorded. As with channeltrons, these capillaries are internally coated with an electron-emissive material allowing ions or electrons striking the inside wall to create an electron cascade. This cascade creates a gain of up to 10^4. The use of two channel plates back to back with one rotated 180° to the other can result in a gain of 10^6. These are sometimes referred to as *chevron MCPs*. These detectors, however, start to saturate when the incoming electron count rate exceeds ~3 × 10^5 cps.

The current pulse generated can then be collected via either

(a) An anode collector
(b) A phosphor screen
(c) A delay line detector (DLD)

An anode collector simply consists of a metal electrode connected up to highly sensitive pulse counting electronics.

Phosphor screens generate photons on electron impact. The use of these, along with an optical lens and camera setup, thus allows for direct imaging of the signal of interest. These are generally referred to as charge-coupled devices (CCDs). Since more than one photon is produced per electron on phosphor screens, further amplification is realized such that the overall gain approaches 10^8.

The DLD directly records the output on a wire in the same manner as an anode encoder, the difference being that the position along the wire at which the signal is recorded allows for the registration of the position at which the initial impact occurred on the MCP. This position is derived by noting the time it takes the signal to travel in both directions along the wire. If the time is the same, the signal is registered as coming from the center of the MCP stack. To obtain a 2-D image, two mutually perpendicular sets of wires are used with each set being a single wire meandering back and forth along a specific axis. These represent a relatively new technology that is fast being implemented in XPS for both pulse counting and imaging (spatial and energy).

Lastly, the amplification factor for EMs is a function of the impinging electrons' $K.E.$ (greater amplification for greater $K.E.$). Amplification factors also vary for the type of EM, the EM age, and the operating conditions used. As a result, quantification is best carried out when the energy analyzer used to energy filter the electron emissions is operated in the CAE/FAT mode (see Section 3.1.5.3). This is realized since this ensures that the energy of all the electrons impinging on the EM lies in the same $K.E.$ range. Postacceleration of the electron emissions can be implemented to enhance EM sensitivities.

3.1.7 Imaging

Imaging in this section refers to the act of mapping the electron emissions with their distribution being representative of the spatial distribution of elements and/or their speciation across the surface of the sample being analyzed. Due to the developments in data storage and instrument design, the area of imaging has experienced significant advances over the last few decades with development still ongoing. As a result, several approaches now exist, as developed by the various different XPS manufacturers. These approaches can be subdivided according to whether the images are collected in

(a) A serial manner (this is covered in Section 3.1.7.1)
(b) A parallel manner (this is covered in Section 3.1.7.2)

Also of interest is the distribution of elements and/or their speciation as a function of depth. The ability to derive depth distributions is discussed in Section 4.3. That being said, one methodology used in acquiring the data necessary in deriving depth distributions, albeit within the sampling depth, employs imaging of electron emissions at different takeoff angles. Since this employs an MCP detector with signals collected in parallel mode, this is covered in Section 3.1.7.2.

3.1.7.1 Serial Imaging   As the name suggests, constructing maps in serial mode requires collecting the data on a point-by-point approach with the image synchronously reconstructed pixel by pixel. There are several approaches in which this can be carried out, these being

(a) Scanning the sample stage

(b) Scanning a finely focused X-ray probe across the surface of interest

(c) Controlling the area from which the electron emissions are collected from

The first approach is only used when images over regions spanning several millimeters in diameter or larger are of interest. As the name suggests, images are acquired by simply moving the sample stage while continuously collecting the electron emissions.

The second approach, otherwise referred to as the *probe defined* mode, is used when images spanning several hundred micrometers in diameter are of interest. This mode requires a monochromotic X-ray source since it is the crystal in these units that focuses the X-ray beam (see Section 3.1.2.2) as illustrated in Figure 3.6a. Scanning is induced by rastering the electron beam used to generate the X-ray beam over the X-ray anode surface. The image is then constructed by synchronizing the electron raster pattern with the timing of the detection electronics. This approach can be applied to instruments using either a CHA or a CMA (see Sections 3.1.5.1 and 3.1.5.2, respectively) since the resolution is set by the X-ray beam diameter. Furthermore, the energy analyzer can be operated in their wide-open mode (lowest-energy resolution) thereby maximizing transmission of the electron emissions and, hence, sensitivity. Full spectra can also be collected for each point (pixel) by scanning the energy filter during imaging. Indeed, there exists highly effective, commercially available instruments that collect images using this approach.

The third approach, referred to as the *lens defined* mode, is again used when images over a region spanning several hundred micrometers in diameter are of interest. This mode controls the area over which the electron emissions are collected and transferred to the detector by the transfer lens/area defining aperture; thus, any X-ray source can be used. An image is constructed by scanning the area from which the electron emissions are collected as illustrated in Figure 3.6b. This is accomplished by ramping the voltages on the deflector units within the transfer lens, which sits prior to the area defining aperture. The need for the transfer lens/area defining aperture thus limits this approach to instruments equipped with a CHA, which can be operated in either the CAE/FAT or CRR/FRR modes (discussed in Section 3.1.5.3, with the former being by far the most common). The sensitivity achievable using this approach is, however, severely limited by the area analyzed relative to the wide-open probe defined mode.

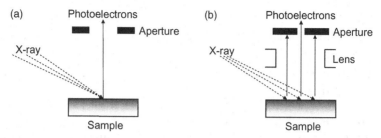

Figure 3.6. Schematic illustrations of imaging in XPS using (a) a small probe X-ray source and (b) the photoelectron transfer lens/analyzer. In the former, the spot size is determined by the X-ray probe diameter and in the latter the transfer lens/area defining aperture.

In any of the above cases, the use of a CHA, along with an array of detectors (multiple channeltrons are typically used as discussed in Section 3.1.6.1) placed along the energy-dispersive axis of the CHA, introduces the possibility of either

(a) Deriving pixel-specific energy distributions, albeit over a narrow energy span and with limited data density (limited by the number of detectors)

(b) Increasing the overall sensitivity by summing the data from neighboring energy channels (those from neighboring detectors)

This capability is realized when operating a CHA in either the CAE/FAT or CRR/FRR modes (see Section 3.1.5.3 with the former being far more common) since electron emissions are separated as a function of $K.E._{XPS}$ along the energy-dispersive axis.

3.1.7.2 Parallel Imaging More recently, a second imaging approach termed *parallel imaging* has been implemented for constructing maps of the distribution of elements and/or their speciation across a surface. This is used for examining areas of less than several hundred micrometers in diameter, with several highly effective commercial instruments available. As the name suggests, electron emissions from the entire field of view are transferred simultaneously without scanning the sample stage, the X-ray source (source defined mode), or the deflectors in the transfer lens (lens defined mode).

Imaging in this mode is accomplished by electrostatically and/or magnetically projecting the electron emissions within some preselected $K.E._{XPS}$ range from the sample's surface through either a CHA or an SMA energy analyzer (different analyzer approaches are used by different instrument manufacturers) and onto a 2-D detector. To do this requires additional lenses placed before and after the energy analyzer to ensure that spatially dispersed signals (electrons) enter and exit the energy analyzer along parallel trajectories, and of course, a 2-D MCP (MCPs are discussed in Section 3.1.6.1).

An example of how images are collected within an instrument equipped with an SMA is shown in Figure 3.7(a). Within this design, the energy of the electrons to be imaged is constrained by a baffle within the SMA placed at the focal point where greatest energy dispersion occurs. Note: As in photon optics, focal points diverge in space

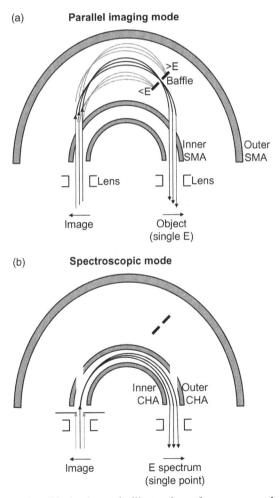

Figure 3.7. Highly simplified schematic illustration of one approach used to acquire (a) images in the parallel microscope mode and (b) spectra in the spectroscopic mode. *Note*: Both information types can be collected sequentially.

with the energy of the electron emissions. This energy filtering must be carried out in order to reduce the effect of chromatic aberrations and thus retain image resolution. The other factor that affects spatial resolution is what is referred to as spherical aberrations (due to beam spread). These are minimized by ensuring the electrons remain close to the instrument's optical axis. Chromatic and spherical aberrations are discussed further in Appendix E. Spectra can then be collected from predefined regions of the sample by deflecting the electron emissions through a CHA encased within an SMA, as is illustrated schematically in Figure 3.7(a).

The advantages of this approach relative to the serial approaches lies in the fact that

(a) Improvements in spatial resolution are noted. Values marginally better than 3 μm are possible in this imaging mode.
(b) Acquisition times for the same sensitivity are reduced. As an example, 256×256 pixel images can be collected in a matter of seconds (this takes substantially longer in serial mode).

One notable disadvantage of the parallel imaging approach lies in the fact that each pixel represents electron emissions at a specific energy (for quantification, the background must also be known). To contend with this, additional images can be collected away from the peak of interest (this provides an estimate of the background signal) or the instrument can be switched back to spectroscopic mode to collect spectra from localized regions of interest (this provides for a more accurate understanding of the peak shape and background and thus a more accurate quantitative data).

As mentioned in Section 3.1.6, there also exists a form of parallel imaging that relays the data necessary to derive depth distributions within the sampling depth in a simultaneous fashion (the ability to derive depth distributions is covered in Section 4.3). This possibility exists since, as illustrated in Figure 3.8, photoelectrons exiting the sample's surface at different takeoff angles can be dispersed along the plane normal to a CHA energy-dispersive plane and mapped on an MCP. As described in Section 4.2.2.2, the takeoff angle controls the depth from which the photoelectrons emanate.

3.1.7.3 *Spatial Resolution*

Spatial resolution describes the ability to separate signals emanating from spatially different locations when exhibiting a different composition/speciation. One definition is the distance over which some signal from the edge of some highly localized

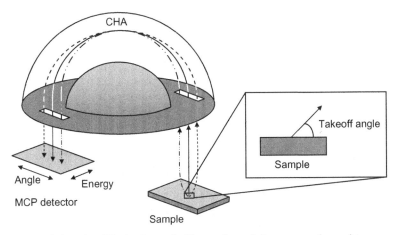

Figure 3.8. Highly simplified schematic illustration of the approach used to map photoelectrons as a function of takeoff angle and $K.E._{XPS}$ in parallel.

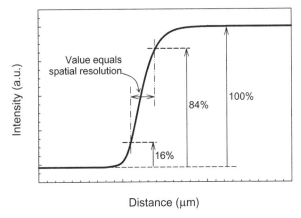

Figure 3.9. Illustration of how spatial resolution is defined (16%–84% definition).

region climbs from 16% of its maximum intensity to 84%, or vice versa, when plotted on a linear scale. This definition, illustrated in Figure 3.9, represents two standard deviations ($\pm 1\sigma$) of the convolution of a Gaussian function with a step function. Other definitions include the distance over which the signal varies from 20% to 80% or even from 10% to 90%.

The spatial resolution that can be acquired on the latest commercially available XPS instruments using a monochromated Al-$K\alpha$ source along with a CHA or SMA when operated in the parallel imaging mode is slightly less than 3 μm. An example of this is displayed in Figure 3.10. Note: The 20%–80% definition was used in this example.

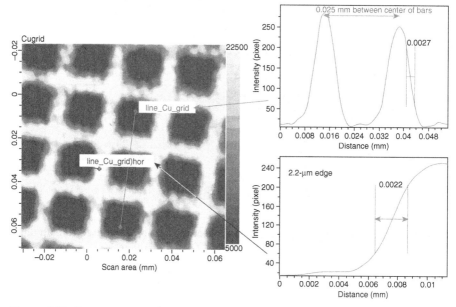

Figure 3.10. Image generated from the Cu-2p photoelectons from a 25-μm pitch Cu grid along with line scan analysis. The 20%–80% definition reveals a 2.2-μm spatial resolution. These data were provided by Kratos Analytical Inc.

This spatial resolution increases to values slightly less than 10 μm when utilizing a CHA with images collected in a serial manner using the probe defined mode (this represents the minimum spot size of a focused monochromated Al-$K\alpha$ source). A further increase to ~15 μm is noted when collecting images in a serial manner using the lens defined mode. Note: What the probe defined mode loses in spatial resolution, it picks up in spectroscopic sensitivity.

Submicrometer scale spatial resolution values in XPS can at present only be arrived at through the use of alternative approaches such as those resulting from the development of energy-filtered photoelectron emission microscopy (PEEM) or, more precisely, X-ray photoelectron emission microscopy (XPEEM). Indeed, a spatial resolution approaching 0.5 μm has been reported when using a purpose-built double CHA along with an Al-$K\alpha$ monochromatic X-ray source (Renault et al., 2007). This spatial resolution is further improved to ~0.1 μm when using a synchrotron source (Renault et al., 2007), with additional improvements expected.

PEEM and XPEEM are techniques belonging to a highly specialized group of methods that examine electron emissions produced through photon irradiation of samples held under high extraction fields (conventional XPS does not use high extraction fields). However, prior to the recent introduction of energy filtering, only the low-energy electron emissions (secondary electrons) could be imaged. Energy filtering has since opened up the possibility of imaging select photoelectron and Auger electron emissions with imaging carried out in parallel mode (see Section 3.1.7.2).

Taking this to the next step, spectroscopic photoemission and low-energy electron microscopy (SPELEEM) is a recently commercialized technique that combines LEEM and XPEEM into a single synchrotron-based instrument. LEEM uses an electron source thereby introducing the added capability of low-energy electron diffraction (LEED), etc., to the pool of characterization techniques that can be applied to the same submicrometer scale volume. Note: Analysis at these dimensions is also made possible through the greater photon flux afforded by synchrotrons (see Section 3.1.2.4).

Such analysis is commonly referred to as *spectromicroscopy*. The term derives from the fact that the term *spectroscopy* is used to describe the act of carrying out chemical analysis alone, while *microscopy* is used to describe the act of imaging alone. The high spatial resolution stems from the fact that under the high fields used, the electrons can be considered as emanating from point sources, these being the photoelectron emitting atoms/ions. This extends to Auger electrons and secondary electrons since both result from the relaxation of the core hole produced.

3.2 SUMMARY

Without the significant technological advances made, going back as far as 1650 at which time the first vacuum pump was demonstrated, XPS would not be the analytical technique it is today. Indeed, XPS would not have been anything more than an academic curiosity if not for the ability of producing UHV conditions (this allowed for the analysis of the surface of interest, as opposed to surface contaminants as well as for the signal of interest to reach the detector). Such vacuums were, however, not attainable until the 1950s. Since then and even now, further advances have and are being made in this active field of research.

There are, in essence, three components to an XPS instrument, these being

(a) The source
(b) The energy analyzer
(c) The detector

The source is the unit from which X-rays are produced/directed onto the sample of interest. Those found in self-contained (stand-alone) XPS instruments utilize an X-ray tube with standard and/or monochromatic sources being the two options. The latter, with Al-$K\alpha_1$ emissions used, are increasingly popular due to the energy and narrow energy spread produced. X-rays of narrower energy spread than this can only be produced in synchrotrons. The energy resolution in XPS is primarily dependent on the source.

Energy filters used in XPS tend to consist of CHAs. These are heavily favored over CMAs due to their superior energy resolution and the fact that when operated under the CAE mode, these provide a constant energy resolution over the entire $B.E._{XPS}$ range of interest. In addition, little variation in $B.E._{XPS}$ values is noted once the sample–analyzer distance is optimized. CHAs also allow for the incorporation of transfer lenses. SMAs, a variant of the CHA, are also used in imaging.

Detectors are comprised of some form of EM. The type used depends on the information of interest. If counts over some large area are required, multiple channeltrons placed along the energy-dispersive axis of the energy filter are the best option. Additional information, that is, imaging of the spatially dispersed and/or angularly dispersed data, can be acquired using an MCP detector when combined with CHA or SMA. Imaging can be carried out in one of several modes, with the parallel mode providing the best spatial resolution.

Highly effective instrumentation is commercially available from several vendors each with attributes of their own, all of which can provide photoelectron signals up to 1×10^6 cps with an energy resolution of 0.3–0.5 eV. The best spatial resolution currently available in self-contained (stand-alone) XPS instruments is presently slightly better than 3 μm. Alternative approaches based around XPEEM can provide submicron resolution with relative ease. These highly specialized approaches allow for true spectromicroscopy.

CHAPTER 4

DATA COLLECTION AND QUANTIFICATION

4.1 ANALYSIS PROCEDURES

If correctly applied, X-ray photoelectron spectroscopy (XPS) has the capability of providing a wealth of information on the elemental composition and speciation of a solid's surface. To correctly apply XPS requires knowledge of

(a) The most effective sample handling procedures for reducing the contamination and/or modifications of the sample surface of interest

(b) The most effective data collection methodologies, along with optimal reference procedures

(c) The degree and type of damage that can be suffered by the sample during analysis

(d) The volume analyzed and how this can be adjusted to suit the need of the respective experiment

(e) Quantification methodologies inclusive of background subtraction routines and potential sources of errors

X-ray Photoelectron Spectroscopy: An Introduction to Principles and Practices,
First Edition. Paul van der Heide.
© 2012 John Wiley & Sons, Inc. Published 2012 by John Wiley & Sons, Inc.

The above aspects, many of which can be sample specific, are covered throughout this chapter. Chapter 5 covers spectral interpretation, while Chapter 6 presents some specific case studies using the methodologies outlined in both Chapters 4 and 5.

4.1.1   Sample Handling

Unlike most analytical techniques, XPS requires little in the way of sample preparation. Indeed, all that is generally required is that

(a) The sample be of a size amenable to that of the instrument
(b) The surface of interest remains clean before and during analysis
(c) The surface of the sample be ultrahigh vacuum (UHV) as well as X-ray compatible

As far as size is concerned, samples should optimally be the size and geometry of a small coin. This is suggested since this allows for ease of handling and is of a sufficiently small surface area to allow for acceptable pumpdown speeds. Note: Access ports on commercially available instruments are also of a limited size as this keeps their cost of purchase and ownership within acceptable levels (larger samples will require larger introduction chambers, larger analysis chambers, greater pumping speeds, etc.).

If the samples are of a few millimeters or larger, they can be affixed to the sample stub or platform using clips of one form or another. If they are smaller than a few millimeters or are in the form of powders, some alternative support is needed. One practice is to use indium foil strips into which the sample is pressed using another piece of indium foil and/or a spatula. The strips are then clipped onto the sample stub/ platform. Another practice is to sprinkle the powder onto double-sided spectra grade sticky tape placed on a flat disk, which is then affixed to the sample stub/platform. Loose powder should always be removed by inverting the stub and/or through the use of pressurized N_2. If nano-sized powders are to be examined, these can be dispersed in some inert solvent and then pipetted onto a substrate and allowed to dry. Note: The above approaches should not be used in high-temperature studies. For these, specialized vendor-specific sample supports should be used. Note: No standardized sample holder type presently exists (all are vendor specific).

An example of a typical sample preparation area is shown in Figure 4.1.

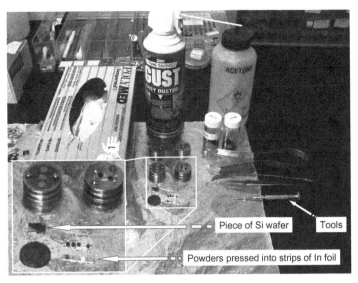

Piece of Si wafer | Tools

Powders pressed into strips of In foil

Figure 4.1. A typical XPS sample preparation area. The circular sample supports (stubs) used in this specific instrument (from Physical Electronics) are magnified. Also shown are indium strips onto which powders have been pressed (this acts as an effective support as discussed in the text).

Cleanliness of the sample prior to and during analysis is imperative since, as illustrated in Figure 1.3, any undesired contamination introduced onto the area of interest will quench the signal of interest. Cleanliness during sample preparation can be assured through the use of gloves and the appropriate precleaned tools (tweezers, screwdrivers, spatulas, sample support holders, etc.). Although latex gloves can be used, polyethylene gloves are preferred (some latex gloves contain silicones, which can segregate to a sample's surface if introduced). Cleaning is usually carried out using acetone, methanol, or isopropyl alcohol (ultrasonic cleaning can also be of use). Tools to mount these samples onto the respective sample stubs or platforms must also be kept clean, likewise for tools needed to ensure the sample is of a size that can fit into the instrument. Under no circumstance should samples be handled with bare hands.

Once mounted, the samples are introduced into an introduction chamber (see Section 3.1.1.1). These are of minimal size to allow for quick pumpdown, with pumpdown times typically in the 10- to 15-minute range for clean samples of low porosity. In the case of highly porous samples, it is advisable to pump these down in the introduction chamber for extended periods of time (overnight prior to analysis may be required). Samples are transferred to the analysis chamber once a pressure in the 10^{-7} Torr range is reached. Introduction chambers thus allow the fast transfer of samples from the atmosphere to the analysis chamber with minimal loss of the UHV conditions required during analysis (days would be required if no introduction chamber existed).

Indeed, one of the reasons UHV is required in XPS is that this controls the readsorption of gas-phase contaminants onto the surface of interest during analysis. Note: This is typically accelerated over regions experiencing X-ray, electron, or ion impact. This requirement, discussed in Section 3.1.1, also places limitations on the sample size (larger samples require longer pumpdown times) and the types of samples that can be examined; that is, the sample must be UHV as well as X-ray compatible. In other words, the sample cannot degas or deteriorate excessively during analysis.

High vapor pressure samples can be analyzed under the UHV conditions required, but only if cooling is implemented (vapor pressures and, hence, outgassing rates scale with temperature). Cooling is typically carried out via the use of an externally situated LN_2 dewar thermally connected to the sample stub or platform while situated within the instrument. Further cooling requires the use of liquid He.

In specific cases, prior removal of undesired surface layers can also be carried through rinsing the sample in specific solvents/reagents prior to placing these in the introduction chamber and/or by heating the sample while in the analysis chamber or in an attached reaction chamber. The latter is preferable since excessive gas evolution can result in rapid deterioration of the vacuum. Heating in the presence of a low-pressure environment of O_2 also can be useful for removing carbonaceous overlayers from thermally stable materials as this readily forms CO_2.

Alternatively, specific surface layers can be removed in the analysis chamber just prior to analysis through sputtering. There are several options with some resulting in less sample damage than others. For example, sputtering with low-energy Ar^+ ions (500 eV or less) can remove surface contaminants with minimal damage. Sputtering with cluster ion beams is another option that has received increased attention due to the reduced damage introduced in removing surface organic layers from organic substrates. These aspects are covered further in Section 4.3.1.2.

4.1.2 Data Collection

Once the sample of interest is within the analysis chamber and the desired vacuum level is reached, prealignment procedures must be initiated to ensure that the sample's surface region of interest is optimally located, that is, that this is the correct distance from the analyzer, and that the area of interest is within the field of view. Since the former can influence the sensitivity and the $K.E._{XPS}$ and $B.E._{XPS}$ scales ($K.E._{XPS}$ and $B.E._{XPS}$ are defined in Section 1.4), this should be optimized for every sample examined.

When analyzing a sample of unknown origin and/or composition, it is advisable to collect spectra over a relatively wide energy range, that is, 0–1000 eV or more (up to the energy of the X-ray beam). This should be carried out under relatively poor energy resolution conditions (e.g., >40-eV pass energy in constant analyzer energy [CAE]/fixed analyzer transmission [FAT] mode when using a concentric hemispherical analyzer [CHA]) with a step density of ~0.5 eV or larger to maximize signal intensities while minimizing analysis times. The spectra, often referred to as *survey spectra*, reveal the most effective emissions for use in quantification and/or subsequent high-resolution studies (high resolution is the most effective when speciation is of interest). An example of a survey spectra collected from an Ag foil sample is shown in Figure 4.2a.

As the same suggests, high-resolution spectra required in speciation studies are collected under high energy resolution conditions (e.g., <40-eV pass energy in CAE/FAT mode when using a CHA) with the step density set at 0.2 eV or less. To offset the longer analysis times required, narrower energy ranges are selected (time is saved by not collecting signals over regions void of signals of interest). To minimize complications, interference-free emissions should be analyzed where possible. An example of high-resolution spectra is shown in Figure 4.2b. These were collected before and after sputtering over the energy ranges signified in Figure 4.2a.

Confusion as to whether a signal arises from photoelectron or Auger electron emission can be removed by adjusting the source (Al-$K\alpha$ to Mg-$K\alpha$) since this results in the movement of all Auger peaks along the $B.E._{XPS}$ scale by an amount equal to the energy difference of the source used (233.0 eV for Al-$K\alpha$ vs. Mg-$K\alpha$ X-rays).

4.1.3 Energy Referencing

XPS derives compositional and speciation information through the energy of the electrons detected. To achieve this requires an energy reference. The only point of reference for electrons in atoms in the gas phase is E_{vac}. Atoms in the solid phase have two reference points, one being E_{vac} and the other being E_F (the second point of reference arises from the fact that conductive solids can be earthed, while atoms/

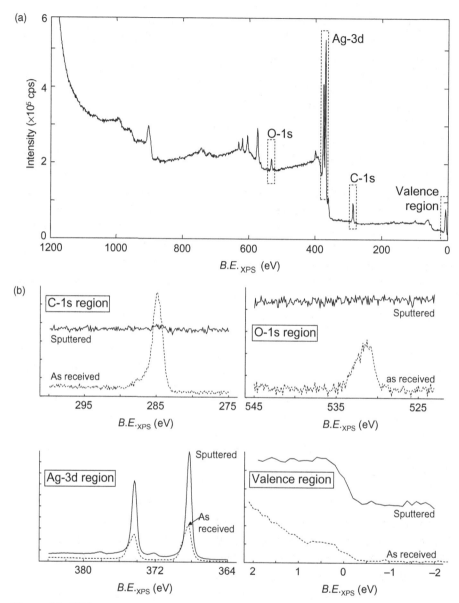

Figure 4.2. XPS spectra from (a) the entire $B.E._{XPS}$ region collected under low-energy resolution conditions from an as-received silver foil, and (b) specific regions (as listed) collected under high-energy resolution conditions from an as-received silver foil before and after sputtering with 1-keV Ar⁺ ions.

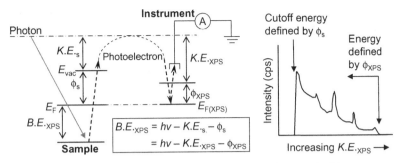

Figure 4.3. The relation between ϕ, E_F, and $K.E.$ of/from the sample (these parameters are followed by the subscript s to denote that they refer to the sample) and the instrument (these parameters are followed by the subscript XPS) is shown on the left. To the right is shown a pictorial representation of their effect.

molecules in the gas phase cannot). The position of E_F defines ϕ; that is, ϕ is the separation between E_F and E_{vac} (see Fig. 2.1) with the value of ϕ defining the minimum energy an electron must have to depart the solid in question (this can be thought of as analogous to the ionization potential of atoms/molecules in the gas phase).

To overcome the possible influence the sample's ϕ (referred to as ϕ_s) can have on the measured $K.E._{XPS}$, and thus the position of E_F as derived from the $B.E._{XPS}$ scale (see Eq. 1.1), instruments are constructed such that their ϕ (referred to as ϕ_{XPS}) values are less than that of almost any sample previously exposed to air. What this does in effect, is replace the dependence on ϕ_s with ϕ_{XPS}. The outcome, illustrated in Figure 4.3, can be understood once realizing that the positions of E_F of any two electrically conductive solids in electrical contact (the sample and instrument) always align with each other, with the outcome (the final ϕ) being that of the predominant source of electrons (the instrument). The value of ϕ_s only affects the position of the high $B.E._{XPS}$ cutoff as also illustrated in Figure 4.3. This results from the fact that if electrons cannot overcome their ϕ_s, they cannot be detected.

If, however, the sample is insulating, this alignment to E_F cannot occur. As a result, the derived $B.E._{XPS}$ scale will shift during electron emission, often uncontrollably to higher $B.E._{XPS}$ values. Methods to control this are discussed in Section 4.1.3.

The value of ϕ_{XPS} (this is needed to solve Eq. 1.1) is most effectively derived by noting the ϕ_{XPS} value needed to ensure the correct $B.E._{XPS}$ value of some well-characterized photoelectron signal, or alternatively defining the energy offset of some well-characterized photoelectron signal from that noted when the value of ϕ_{XPS} is set to zero. To ensure the entire energy scale is calibrated (the same ϕ_{XPS} value should apply

TABLE 4.1 Commonly Used Photoelectron Signals from Sputter-Cleaned Surfaces for Energy Referencing in XPS

Sample	Peak	$B.E._{XPS}$ (eV)	FWHM (eV)
Cu	Cu-2p$_{3/2}$	932.66	≥0.7
Ag	Ag-3d$_{5/2}$	368.26	≥0.6
Ag	E_F	0.00	N.A.
Au	Au-4f$_{7/2}$	83.98	≥0.8

over all energies), as opposed to one specific region, several peaks that span a significant portion of the energy range accessible should be used. Recall: Since the sample–analyzer distance also affects $K.E._{XPS}/B.E._{XPS}$ scales, this should also be optimized.

The narrow core-level Cu 2p$_{3/2}$, Ag-3d$_{5/2}$, and Au-4f$_{7/2}$ peaks from the respective elemental samples are most commonly used for the calibration of XPS instruments. Their $B.E._{XPS}$ values and peak widths (full width at half maxima [FWHM]) are listed in Table 4.1. The sharp drop-off around E_F in the Ag valence region can also be useful since this should align with zero on the $B.E._{XPS}$ scale. Note: The midpoint should be set to zero as shown in the valence band spectra in Figure 4.2b to account for instrument broadening effects.

Such calibration should be carried out under higher-energy resolution conditions and should only take place after all surface oxides have been removed. This is illustrated in Figure 4.2b in which spectra from the Ag-3d$_{5/2}$ and Ag valence regions are shown. Surface oxides should be removed since this will ensure the removal of any potential sources of error while enhancing photoelectron signal intensities. The greater intensities result in part from the removal of the surface overlayer as illustrated in Figure 1.3.

During the analysis of insulators (see Section 4.1.3), there also exists an additional peak that can be useful in ensuring accurate calibration of the $B.E._{XPS}$ scale. This is the C-1s peak arising from adventitious carbon. Adventitious carbon arises from the adsorption of aliphatic hydrocarbons from the atmosphere, which includes that present under UHV (see Section 3.1.1). These are useful since they can provide an effective internal reference check but only if the C-1s $B.E._{XPS}$ from these adsorbates on the respective surface is well-known to within 0.1 eV. Note: These can vary by as much as an electronvolt or more from the customarily used value of 284.7 ± 0.1 eV. This variation is believed to stem from surface dipole effects experienced by the adventitious C overlayer. Of note is the fact that X-ray irradiation of a sample enhances C adsorbtion rates relative to unirradiated areas. Although not significant, the result will reflect the state of the analysis chamber.

4.1.4   Charge Compensation

If the solid being analyzed is insulating, positive charge will accumulate over the area from which electron emission occurs (this is not observed in conductors since electrons from a grounded terminal acts to compensate for this charge loss). Any positive charge buildup has the detrimental effect of reducing the $K.E._{XPS}$ of any subsequently emitted electrons, which, in turn, has the effect of increasing their apparent $B.E._{XPS}$. If this charge buildup is not too extreme, any symmetric peaks will become asymmetric in shape and will move to higher $B.E._{XPS}$ values (lower $K.E._{XPS}$ due to the greater attraction to the surface experienced). If the charge buildup is more extreme, this will result in complete loss of signal as illustrated in Figure 4.4.

Charge buildup can be compensated for through coirradiating the sample with a low-energy electron flood gun (10 eV or less). Low-energy electrons are preferred since charging of the sample surface effectively deflects the low-energy electrons to the affected area thereby setting up a self-compensating mechanism. This is, of course, on the assumption that the incoming electron beam current density (controlled by space charge effects) exceeds that of the charge buildup. This charge compensation process can be facilitated by the immersion of the sample in a weak magnetic field since this also acts to direct the electrons to the area of interest. Higher-energy electrons can be used

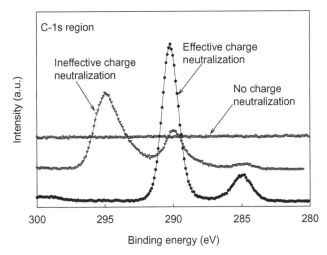

Figure 4.4. Spectra collected over the C-1s region from LiAlO$_2$ powder analyzed under different charge neutralization conditions (CHA set to low-pass energy). The peak at 290.2 eV represents C in the form of carbonates, while that at 285 eV represents that of the adventitious C overlayer.

in charge compensation but only if the electron current matches the charge buildup and the beam is directed at the area affected. This procedure, used in older instruments, can be difficult to set up and has the detrimental side effect of introducing greater sample damage.

Effective charge compensation will result in symmetric peaks displaying $B.E._{XPS}$ values within a few eV of their true value. The disparity is then typically corrected for during data processing.

Other methods used to ease charge compensation include

(a) Coirradiating the sample with both low-energy electrons and ions. Low-energy inert ion beams are used since this agitates the surface atoms (destroys the band structure) while minimizing sample damage. Damage is minimized since the ion impact energy is below the sputtering threshold energy (~30 eV per atom within the incoming ion).

(b) Using an X-ray source with a lower flux density, for example, a standard X-ray source or unfocused monochromatic source. Although the effectiveness is limited when used alone, it can ease charging when combined with one of the other methods described herein. Charging is minimized as a result of

    (i) The reduced removal of charge from the area being analyzed (this is in the form of photoelectron, Auger electron, and secondary electron emissions)

    (ii) The increased flux of low-energy electrons (primarily secondary electrons) from surrounding areas (that outside of the analyzed volume), which will then be attracted to the areas experiencing greater charge buildup

(c) Placing an earthed conductive grid (tungsten grids are common) or a conductive tape (aluminum or copper tape) in close proximity to the area analyzed. This works particularly well when a defocused X-ray beam is used for reasons described above (see point ii). Note: If care is not taken, this can introduce additional signals (those from W, Al, Cu, and the adhesive).

(d) Coating the sample with a conductive layer, for example, carbon or gold prior to analysis. Note: This can severely quench the signal of interest while introducing additional signals (those from C, Au, and any impurities that may be introduced during the application of the conductive layer).

(e) Pressing the sample into indium foil. This applies particularly well to powdered samples (this also acts as a support as discussed in Section 4.1.1) and when combined with one of the other methods described above. Note: This can introduce additional signals if care is not taken (those from indium).

(f) Heating the sample during analysis. This promotes electrons into the conduction band of large bandgap materials when temperatures greater than ~400°C are used. This method should, however, only be applied to thermally stable precleaned samples.

4.1.5   X-ray and Electron-Induced Damage

One of the common misconceptions associated with XPS is that it is a nondestructive technique. There are many instances where damage to the sample being analyzed is noted during analysis. This stems from the fact that X-ray irradiation can induce

(a) Desorption of surface-bound species
(b) Reduction of atoms/ions within the area analyzed
(c) Phase changes within the sample being studied
(d) Electric field-induced migration of specific elements in specific samples

This is particularly evident when analyzing nonconductive materials and/or materials with loosely bound adsorbates. Indeed, X-ray irradiation of such samples will result in

(a) The removal of charge from the area being analyzed
(b) The perturbation of the electronic structure of the sample under study

The former is the primary culprit. Damage resulting from the removal of charge can be understood if

(a) The photoelectron emitting atom/ion can exist in multiple oxidation states.
(b) The sample under study contains
    (i)   Loosely bound adsorbates
    (ii) Closely situated phases
    (iii) Ions known to migrate under electric fields, that is, Li^+ and Na^+

These effects cannot be compensated for through coirradiation of the sample with an electron beam. Indeed, energetic electron beam irradiation often results in even greater surface damage. This arises since electrons deposit a greater fraction of their energy over the outer surface region than photons of the same energy, a fact resulting from the reduced path lengths of electrons relative to photons (path lengths are discussed further in Section 4.2.2.1). This also explains why techniques such as Auger electron spectroscopy (AES) and electron energy loss spectroscopy (EELS) typically result in greater sample damage than XPS.

4.2 PHOTOELECTRON INTENSITIES

The intensities of photoelectrons are dependent on a number of parameters that can be subdivided into three basic groups as portrayed in Spicer's three-step model (see Berglund and Spicer, 1964). These group-specific parameters include

(a) Those dependent on the physics of photoelectron production
(b) Those that describe the interaction of the photoelectrons produced with the surrounding solid
(c) Those dependent on the instrument geometry

The parameters dependent on the physics of photoelectron production are generally encompassed within the *photoelectron cross section* or probabilities. These cross sections describe the likelihood for photoelectron production from a specific core level of a specific element under the irradiation of photons of a specific energy. What is most important here is that these probabilities do not vary with the chemical environment of the photoelectron emitting atom/ion, including oxidation state. Such cross sections are covered in Section 4.2.1.

The parameters that describe the interaction of photoelectrons with the surrounding solid-state environment are those that result in a loss of photoelectron intensity. These parameters, generally represented by the photoelectron's *inelastic mean free path* (IMFP), describe the mean distance an electron of a specific kinetic energy ($K.E.$) can travel in a particular solid before it interacts and loses energy to its surroundings. If this energy loss is greater than the energy resolution used, the signal will become part of the background. This arises from the fact that the energy loss experienced is typically of a nondiscrete nature. These aspects are discussed further in Section 4.2.2.1.

The parameters dependent on instrument geometry/experimental setup describe such things as

(a) The detection probability of a photoelectron leaving the solid's surface

(b) The dependence of relative photoelectron intensities on the X-ray source to the electron emission collection angle

(c) The dependence of photoelectron intensities on the takeoff angle and crystal orientation, if applicable

The first factor is described by the instrument's transmission function. Although this will depend on the analytical conditions used, this will remain constant for a particular set of analytical conditions.

The second factor, referred to as the *angular asymmetry factor* (Reilman et al., 1976), describes variations in photoelectron intensities on the unpolarized photon (X-ray) energy and angle. This is element and level specific since the X-ray source to electron collection angles are fixed; this instrument-specific dependence remains constant (this is transparent to the instrument operator since it is contained within the quantification software). It is this dependence that explains why intensity ratio variations are noted in the analysis of the same sample on instruments of different geometry.

The third factor describes experimentally introduced effects, for example, the dependence of photoelectron intensities on the takeoff angle. This follows a roughly cosΘ distribution where Θ is taken at 90° along the surface normal (photoelectron intensities peak in this direction). Since all photoelectrons exhibit the same distribution, this too can be assumed constant under the conditions used.

In addition, electrons traveling within a crystalline solid can suffer diffraction effects (this is the basis of low-energy electron diffraction [LEED] as discussed in Appendix H). This arises from the fact that the de Broglie wavelength (λ) of low-energy electrons is of the order of atomic dimensions; that is, 20-eV electrons have a λ of 0.27 nm. If satisfying the Bragg diffraction criteria (see Section 3.1.2.2), constructive interference will be noted along specific crystallographic directions. In other words, photoelectron intensities will be enhanced along specific emission angles. Although the intensity spikes introduced into the angular distribution can be useful for studying structure (basis of *photoelectron holography*, which can provide element-specific surface crystallographic data), they will introduce errors in XPS quantification if not accounted for or screened out.

Also of note is the fact that electrons suffering elastic scattering can emanate from depths greater than that implied by its IMFP if analysis is carried out at off-normal angles. Analysis volumes are discussed in Section 4.2.2.

4.2.1 Photoelectron Cross Sections

The probability that a photoelectron will be ejected from an atom/ion following photon irradiation is defined as the photoelectron cross section. These vary from element to element and from level to level as illustrated in Figure 4.5.

Photoelectron cross sections do, however, remain constant for a particular set of analytical conditions including X-ray energy. Indeed, this is the reason for the relative ease of quantification in XPS when using

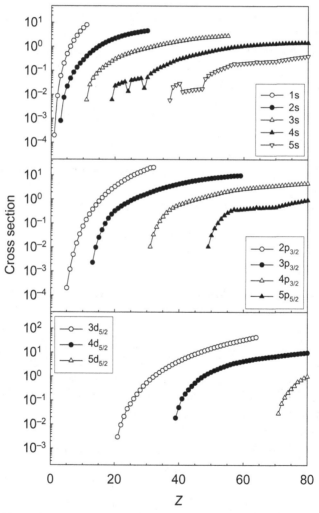

Figure 4.5. Photoelectron cross sections from various levels (stationary states) as a function of atomic number under Al-$K\alpha$ irradiation (Scofield, 1976).

Al-$K\alpha$ and Mg-$K\alpha$ X-rays (this is discussed further in Section 4.2.4). Since these have been tabulated for all elements from Li–U under Al-$K\alpha$ and Mg-$K\alpha$ irradiation by Scofield (Scofield, 1976), these are often referred to as *Scofield cross sections*. Note: If other sources are to be used, the sensitivity factor associated with the respective source must first be derived before accurate quantification can be arrived at.

4.2.2 The Analyzed Volume

The volume analyzed varies with the *K.E.* of the photoelectron and to a lesser degree, the solid itself (these dependencies are discussed in Section 4.2.2.1). If the solid contains at least two different elements, the *K.E.* of their respective photoelectrons generally also differ (if the same, the signals will overlap). This dependence on *K.E.* must be accounted for even in the analysis of homogeneous samples; otherwise, the element with the deeper analyzed volume will appear to have a higher concentration than it actually does. To complicate matters, most solids exhibit some degree of inhomogeneity whether in the form of adsorbed overlayers or some surface-induced segregation or diffusion process.

The primary factors controlling the volume analyzed are

(a) The distance a photoelectron can travel within a solid before losing energy, that is, before suffering inelastic collisions as described by the IMFP.

(b) The angle as specified relative to the surface from which the photoelectrons are collected. This angle, referred to as the takeoff angle (Θ), is discussed in Section 4.2.2.2.

The distance a photoelectron can travel in a solid is defined by elastic and inelastic scattering with the latter being the dominant factor. Elastic scattering describes collisions in which no energy is exchanged, whereas inelastic collisions describe collisions in which energy is exchanged. Since inelastic scattering is of a statistical nature, average or mean values are used. Several definitions, however, are in use. These are

- *The IMFP (or λ_{IMFP})*: This defines the average distance traveled by an electron of a specific energy within a particular single-layered homogeneous amorphous solid between two successive inelastic scattering events.

- *The Attenuation Length (AL)*: This defines the average distance traveled by an electron of a specific energy within a particular multilayered and/or heterogeneous solid (may be amorphous or crystalline) between two successive inelastic scattering events, with elastic scattering included.

- *The Information Depth*: This defines the depth below the surface from which a specified percentage of the photoelectrons emanates.
- *The Sampling Depth*: This equates to the depth from which 95.7% of all photoelectrons emanate. It is equivalent to $3\lambda_{IMFP} \cos\Theta$.
- *The Escape Depth*: This defines the distance normal to the surface from which 61.7% of the original photoelectron population originates. It is defined as $\lambda_{IMFP} \cos\Theta$.

The information depth and sampling depth are those that should be applied when specifying the concentrations of elements within a surface or subsurface layer. These, as well as the escape depth, most precisely describe the statistical nature of the scattering events responsible, that is, that the population of electrons of a specific energy emanating from a solid decreases exponentially with their depth of origin (61.7% and 95.7% translate to a drop in intensity of e^{-1} and e^{-3}, respectively, as described in the following section and in Appendix D). As a result of this, the probability, albeit small, that electrons of some specific energy can escape from a much greater depth than expected based on their IMFP is apparent.

4.2.2.1 *Electron Path Lengths*

The probability of electron–electron interaction is large when compared to photon–electron interaction. This explains why techniques that record electron emissions such as XPS and AES are surface specific while techniques that sample accompanying photon emissions such as X-ray fluorescence (XRF) and energy-dispersive X-ray analysis (EDX) are not. Note: A core-level photoelectron losing a nondiscrete energy more than that specified by the energy resolution used becomes part of the background.

Electron–electron interaction probabilities are most commonly derived through empirical or semiempirical methods. These reveal that for electrons with *K.E.* values greater than ~100 eV (path lengths increase with increasing *K.E.* above 100 eV), interaction probabilities follow an exponential function. The path length of electrons relative to the initial direction of travel can thus be approximated using the *Beer–Lambert law*, a law restated by Lambert and later reformulated by Beer from the original finding by Bouguer (1729). This provides an average value that follows a Poisson distribution (such distributions are covered in Appendix D).

The path length, and hence the parameters AL and λ_{IMFP}, can thus be defined as the distance over which a signal drops in intensity by one order of magnitude when plotted on the natural log scale (e^{-1}), or a 61.7% loss when the intensity is plotted on a linear scale. A drop in intensity of 85.6% therefore equates to e^{-2} and 95.7% to e^{-3}. If non-normal takeoff angles are used in defining λ_{IMFP}, a 61.7% loss translates into the Escape depth ($\lambda_{IMFP} \cos \Theta$) and a 95.7% loss translates into the sampling depth ($3\lambda_{IMFP} \cos \Theta$).

The difference between AL and λ_{IMFP} (the definitions are listed in Section 4.1.5) stems from the fact that the former includes the effects of elastic and inelastic scattering while the latter only accounts for inelastic scattering. Elastic scattering is included since, although not in itself inducing any change in energy, it can increase the apparent depth from which electrons can appear to emanate when layers of different electron densities are studied at off normal takeoff angles.

Although numerous empirical and semi-empirical expressions have been formulated to approximate the value of λ_{IMFP} (research is ongoing), the most accurate expression developed thus far is the TPP-2M relation. TPP is short for Tanuma, Penn, and Powell, who developed this relation (Tanuma et al., 1993). This relation represents a modified form of the Bethe equation (Bethe, 1930), with the modification taking the form of the inclusion of free electron effects that are also responsible for the plasmon loss features (these are discussed in Section 5.1.1.3.2.3).

One form of the TPP-2M relation appears as:

$$\lambda_{I.M.F.P} = K.E./\{E_{pl}^2[\beta\ln(0.191 \cdot \rho^{-0.5}.K.E.)-(C/K.E.)+(D/K.E.^2)]\} \tag{4.1a}$$

with

$$E_{pl} = 28.8(N_v \cdot \rho/M)^{-0.5} \tag{4.1b}$$

$$\beta = -0.0216 + 0.944/(E_{pl}^2 + E_g^2)^{0.5} + 0.000739\rho \tag{4.1c}$$

$$C = 1.97 - 0.91(E_{pl}^2/829.4) \tag{4.1d}$$

$$D = 53.4 - 20.8(E_{pl}^2/829.4) \tag{4.1e}$$

where $\lambda_{I.M.F.P}$ is in units of Å, ρ is the mass density of the solid in g/cm^3, M is the atomic weight of the substrate in units of g/mol (these are listed in Appendix A), E_g is the bandgap energy in eV, and E_{pl} is the free electron plasmon energy with N_v defined as the number of valence electrons per atom. The parameters β, C, and D in Equations 4.1c–e are

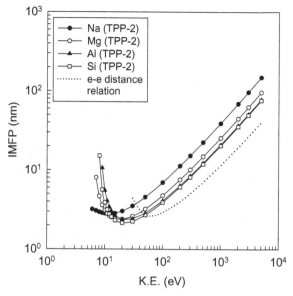

Figure 4.6. Inelastic mean free paths (distance) of electrons in the solids listed as a function of kinetic energy as derived using semiempirical methods (the TTP-2M relation presented in Eq. 4.1a–e) and theoretical methods (see Eq. 4.2).

empirically defined fitting parameters. Note: Equation 4.1c can be rewritten using a $\rho^{0.1}$ factor.

The values of λ_{IMFP} for <10 to 10,000 eV electrons within the elemental solids of Na, Mg, Al, and Si as approximated by the TPP-2M relation are shown in Figure 4.6. These were chosen due to the progression in N_v of these elements, their density variations, and the fact that Si is the only element from this group that has a bandgap. As can be seen, these exhibit similar trends irrespective of the solid studied. This arises from the fact that λ_{IMFP} depends primarily on the free electron density, which, as noted in Equation 4.1b, is a function of N_v, ρ, and M, all of which affect the inelastic scattering occurring at differing $K.E.$ values. Sample and instrument geometry dependent effects must be included in developing expressions for AL.

From a theoretical standpoint, the distance traveled by an unbound electron (the distance between inelastic scattering events) is dictated by electron–electron and electron–proton collisions. The latter can, however, be disregarded over the energy range of interest in XPS since this has minimal effect. In the case of solids, the incoming electrons face the greatest probability of interacting with electrons closest to the Fermi edge through inelastic processes, hence the reason for the inclusion of N_v in Equation 4.1b.

Since free electrons in solids can be viewed as a plasma, the probability for electron–electron interaction can also be approximated from the electron density, and hence the mean electron–electron distance (r_s). The mean free path $(\lambda_{M.F.P})$ can then be shown (Ibach, 1977) to scale as

$$\lambda_{M.F.P} = 1/\{\sqrt{3} \cdot r_s^{3/2}[(a_o \cdot R)/K.E.)] \cdot \ln[(4/9\pi)^{2.3} \cdot r_s^2 (K.E./R)]\}, \quad (4.2)$$

where a_o equals 0.529 Å, R equals 13.6 eV, and r_s is in units of Bohr radii (equals 0.529 Å). The distance traveled as a function of the electron $K.E.$ using an arbitrarily set value of 1.5 for r_s is plotted in Figure 4.6. This compares most closely with that derived for Au via the TPP-2M relation.

Lastly, the fact that free electrons can be considered a plasma reveals why electron emissions from all solids appear to exhibit similar mean free path versus $K.E.$ trends, which has been interpreted in the form of a *universal curve* (Seah and Dench, 1979). This however, is only an approximation since, as made evident in Figure 4.6, electrons of the same energy in different solids can exhibit different path lengths.

4.2.2.2 *Takeoff Angle* The analyzed volume also depends on the takeoff angle (Θ) used in data collection (defined relative to the sample surface in XPS). This can be understood since adjusting the angle of the sample surface with respect to the photoelectron collection optics adjusts the distance a photoelectron must travel within the solid if coming from the same depth normal to the surface (smaller takeoff angles require photoelectrons from the same depth to travel larger distances). This effect scales as the cosine of the takeoff angle and explains the cosΘ dependence noted in the sampling depth and the

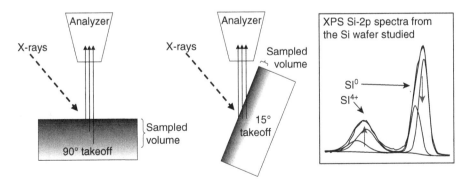

Figure 4.7. The middle and left images show how the takeoff angle controls the volume analyzed (gray area). The spectrum to the right illustrates how this alters the Si-2p peaks from a silicon wafer bearing a 1-nm surface oxide (see vertical arrows). A 45° takeoff angle was used in acquiring this spectrum.

escape depth relations (see Section 4.2.2). An example of how the takeoff angle controls the analysis volume is shown in Figure 4.7.

Figure 4.7 shows the spectra (Si-2p region from a Si wafer bearing a ~1-nm native oxide) and how this is modified; that is, the Si-2p oxide peak (that at 103.7 eV) increases in intensity relative to the bulk peak (that at 99.4 eV) as the takeoff angle is decreased. This will have the effect of decreasing the Si concentration derived since the oxide region now represents a greater fraction of the analyzed volume.

4.2.3 The Background Signal

One of the primary problems in quantifying XPS spectra lies in dealing with the background signal. This is most evident in spectra collected from transition metals. The background signal arises from a number of sources with the most prevalent being inelastic scattering of electrons. As discussed in Section 4.2.2, the nondiscrete energy loss resulting from inelastic scattering (energy lost by electrons still within the solid to its surroundings) produces a broad background to the lower $K.E._{XPS}$ side (higher $B.E._{XPS}$) of the peak of interest.

There are three commonly used subtraction routines for describing and removing the background from peaks of interest (Fig. 4.8). These are the

(a) Linear routine

(b) Shirley routine

(c) Tougaard routine

The linear routine, as the name suggests, assumes a linear function drawn between two points located on either side of the photoelectron peak of interest. Although a very simple approach, it can provide acceptable results. Due mainly to computational limitations at the time,

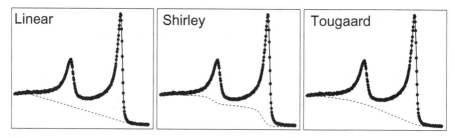

Figure 4.8. Examples of the three primary background subtraction routines used in XPS. These are represented by the respective dashed lines underlying the Fe-2p spectra from a presputtered iron foil (CHA set to low-pass energy).

this was the most heavily used method in the early days of XPS (Briggs and Seah, 1990).

The Shirley routine (Shirley, 1972b) assumes an S-shaped background per peak, with the form defined by the intensity trend noted on the lower and higher $B.E._{XPS}$ side of the respective photoelectron peak and the peak intensities themselves. Although this is a subjective weighted average approach (an ad hoc method), its relative simplicity and acceptable accuracy, assuming fixed limits are applied, have made this the most popular routine presently in use. Note: Quantification requires that a repeatable systematic approach be used (this provides the needed precision). Errors (in accuracy) are then accounted for through use of appropriate sensitivity factors.

The Tougaard routine (Simonsen et al., 1999; Tougaard, 1988) is the only method that attempts to calculate the background, that is, attempts to simulate the inelastic scattering events active, albeit over a limited energy range. As a result, this is the least subjective but the most complex of these three approaches. In an effort to reduce the complexity (as well as the necessity to fully understand all instrumental factors), a *universal loss function* has been derived, which allows the *fitting* of this background over the relatively narrow energy range typically of interest.

A less commonly used but highly accurate approach is to derive the background obtained from reflected electron energy loss spectroscopy (REELS) spectra collected within the same instrument. This method, drawn from AES, requires an energetic electron gun of sufficient energy to analyze the region of interest.

4.2.4 Quantification

One of the strengths of XPS lies in its ability to quantify the recorded signals with relative ease and without the need for reference materials. Under such conditions, an accuracy of no better than 5%–10% can be expected with errors up to 30% arising if

(a) The background is not effectively accounted for.

(b) Inaccurate electron IMFP values are used.

(c) A significant concentration variation exists within the region analyzed. This inhomogeneity may be due to

    (i) The presence of microscopic inclusions

    (ii) The presence of macroscopic regions of different compositions

    (iii) Thin films (less than the sampling depth) of different compositions.

Note: Quantification in XPS assumes homogeneous amorphous-like mixtures over the volume analyzed with no diffraction effects present. The accuracy can be improved to better than 5% if reference materials displaying a similar composition are analyzed with the sample of interest, as this removes/cancels out most background, IMFP, and/or concentration gradient-induced errors.

Quantification without the aid of reference materials or any prior knowledge of the sample is possible since the intensity of a particular photoelectron peak can simplistically be expressed in the absence of elastic scattering events as (see Briggs and Seah, 1990 for a more complete discussion)

$$I = J \cdot c_a \cdot \alpha_{pc} \cdot K_f \cdot \lambda_{IMFP}, \tag{4.3}$$

where J represents the X-ray flux striking the analyzed area; c_a is the concentration of the photoelectron emitting atom/ion within the sampled volume (a is the element of interest); α_{pc} is the photoelectron cross section; K_f encapsulates all instrument factors such as the transmission function; and λ_{IMFP}, not to be confused with λ (wavelength), is the IMFP of the photoelectron (this approximates the volume analyzed).

The parameters α, K_f, and λ_{IMFP} are typically transparent to the analyst since they are accounted for within the instruments software, and J drops out since this remains constant during analysis. Due to the intrinsic energy spread of a photoelectron signal (core holes affect the FWHM of photoelectron peaks as discussed in Section 5.1.1) and the finite energy resolution of the instrument used (see Section 3.1.5), I in Equation 4.3 must be taken as the area bounded by the photoelectron peak and the background (the background is discussed in Section 4.2.3).

Although proven effective, this relation can suffer quantification errors in the analysis of homogeneous samples (particularly those containing transition metals) if the background signal is ineffectively modeled (the background is covered in Section 4.2.3). In addition, final state effects in the form of shake-up and shake-off processes as well as plasmon formation will rob the main photoelectron peak of intensity while introducing greater difficulty in background subtraction.

In order to reduce the resulting quantification error, a systematic approach should be used with sample-specific sensitivity factors applied (see Eq. 4.3). During the analysis of inhomogeneous materials, additional errors can be introduced through the assumption that the volume analyzed is homogeneous and/or if λ_{IMFP} is ineffectively described (this is covered in Section 4.2.2.1).

Improved accuracy can be realized when λ_{IMFP} in Equation 4.3 is replaced by the AL (see Section 4.2.2). This is particularly effective when photoelectrons are collected at off-normal angles from multilayered structures, that is, metals with oxide overlayers, a fact realized since elastic scattering can allow the escape of electrons from deeper layers than that indicated by λ_{IMFP}. To apply these, however, requires that the value of AL be derived or, alternatively, sample-specific sensitivity factors be obtained from matching reference materials (as described in Equation 4.4, the AL is accounted for in the latter).

In this first principle approach the terms α_{pc}, K_f, and λ_{IMFP} are encompassed into what is referred to as the *sensitivity factor* (F). The factor J is omitted since this remains constant during a particular analysis. The sensitivity factor is thus specific to the instrument used, as well as the element and level the photoelectron emanated from. This can be modified according to any extraneous background signals that may be present, that is, it can be adjusted to provide a known concentration following the analysis of a reference material. The concentration of the pertinent element is then derived from the sample of interest by dividing the intensity (area) of the pertinent level by its specific F; that is,

$$c_a = (I_a/F_a)/((I_a/F_a)+(I_b/F_b)+\ldots)\times 100, \qquad (4.4)$$

where the subscripts, a, b, and so on, represent different elements. In either case, all major and minor elements must be accounted for; otherwise, the values derived will only represent relative ratios. When this is not possible, that is, when there exists a substantial hydrogen content as in hydrides, one can only attempt to renormalize the data in a systematic manner on the basis of values derived via some other technique, or those expected (stoichiometry).

Note: A single standardized approach in quantification is not used in the software supplied by the different XPS instrument manufacturers.

4.3   INFORMATION AS A FUNCTION OF DEPTH

In many instances, it is desirable to not only obtain information on the composition and speciation of the elements present at the outer surface of a solid but also to understand how these may vary as a function of depth (as indicated in Section 1.1, there are numerous reasons as to why such variations may exist). If the resulting layers are spatially homogeneous, or at least homogeneous within the area analyzed, several methods exist for extracting this information via XPS. These are discussed henceforth.

4.3.1   Opening up the Third Dimension

As discussed in Section 4.2.2, electrons can only travel a short distance within a solid before experiencing inelastic energy loss. As an example, 95.7% ($3\lambda_{IMFP}$) of electrons with $K.E.$ values less than 1 keV will travel less than ~6 nm in Si (see Fig. 4.6) with 61.7% interacting within one-third of this distance. Information on the depth distribution of elements within this region of the solid can be attained by utilizing angle-resolved X-ray photoelectron spectroscopy (AR-XPS) or energy-resolved XPS. These possibilities stem from the fact that the sampling depth is a function of both Θ and λ_{IMFP} with the latter being $K.E.$ and sample dependent. This is discussed in Section 4.3.1.1. Information on the depth distribution of elements over depths greater than this can only be attained through cosputtering of the area of interest, that is, sputter depth profiling, or through point-by-point analysis carried out across the cross section of the sample. The former is best applied if the depth scale of interest is less than ~10 μm, while the latter is best applied when the depth scale of interest is greater than ~10 μm. Sputter depth profiling is discussed in Section 4.3.1.2.

4.3.1.1   AR-XPS and Energy-Resolved XPS   If the depth of interest is equal to or less than the maximum sampling depth ($3\lambda_{IMFP}$), then it is possible to extract information on the depth distribution of elements through the use of either

(a) *AR-XPS*: This is realized since the sampling depth scales with $\cos\Theta \cdot 3\lambda_{IMFP}$, where Θ is the collection angle relative to the surface (see Fig. 4.7). Analysis is best carried out using a CHA due to their smaller collection solid cone relative to a cylindrical mirror analyzer (CMA). This can by carried out by

   (i) Tilting the sample with respect to the analyzer. Although analysis at multiple takeoff angles is carried out in a serial manner, an automated routine can be set up to save time and effort.

   (ii) Using specifically designed optics such as that illustrated in Figure 3.8. This form is referred to as parallel angle-resolved X-ray photoelectron spectroscopy (PAR-XPS).

(b) *Energy-Resolved XPS*: This is realized since λ_{IMFP} varies with the K.E. of the electrons measured (see Fig. 4.6). This can be observed if spectra from two different core levels from the same element produced by the same X-ray source are compared, or from spectra of the same core level produced by different X-ray sources, that is, Al-$K\alpha$ and Ag-$L\alpha$ sources. Energy analysis can be carried using a CHA or CMA.

An example of these possibilities is illustrated in Figure 4.9, in which both the Ba-$3d_{5/2}$ and 4d photoelectrons are recorded from a single-crystal BaTiO$_3$ substrate. Note: Surface-bound barium exhibits higher B.E.$_{XPS}$ values than subsurface barium. Since the λ_{IMFP} of the 4d photo-electrons is greater than that of the $3d_{5/2}$ photoelectrons, the contribution from surface-bound Ba is less evident in the 4d spectra. Increasing

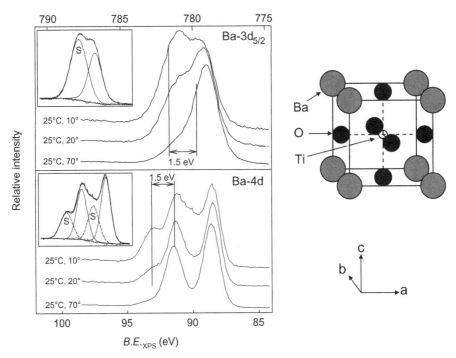

Figure 4.9. Photoelectron spectra over the Ba-$3d_{5/2}$ and Ba-4d regions from clean BaTiO$_3$ collected at the takeoff angles listed (CHA set to low-pass energy). Curve fitting of the spectra collected at 10° is shown in the respective insets to reveal the surface (denoted by s) and subsurface components. To the right is shown the crystal structure of BaTiO$_3$ (Jona and Shirane, 1993).

the takeoff angle further reduces this surface component in line with
Equation 4.1a–d.

As a result of these dependencies, one can define

(a) The order of chemically distinct layers in a multilayered
    structure
(b) The thickness of the topmost layer, assuming a smooth interface
    exists

Assuming the structure is not too complex, the order of the elements
present, or even their speciation, can be defined from the relative intensities of the respective electron emissions with takeoff angle. Plots
derived from this information are typically referred to as relative depth
plots (RDP).

There are, however, instances where depth distribution of the concentration of the elements present, or even their speciation, is useful.
This can be accomplished using the maximum entropy concept to
depths extending up to $3\lambda_{IMFP}$ (those within $2\lambda_{IMFP}$ will be more accurate). This method derives an error between the derived profile and
that implied by the experimental data along with an entropy term using
the RDP as a first step. Subsequent steps then attempt to minimize the
error while maximizing the entropy (third law of thermodynamics).
Vendor-specific recipes can be effective in this regard. If not available,
an in-depth understanding of the maximum entropy method is needed.
Comparison with medium-energy ion scattering (MEIS)-derived depth
profiles is also advisable since this is the only other nondestructive
method presently available for supplying elemental depth profiles over
sub-10-nm regions (MEIS is discussed in Appendix F).

The thickness (T) of some overlayer can be accurately derived via a
modification of the Beer–Lambert law. As an example, the surface
oxide formed on Si can be derived from the relative photoelectron
intensities (I) of the Si^0 and Si^{4+} peaks in the Si-2p spectra to better
than 0.05 nm (Mitchell et al., 1994) via the relation

$$T = \lambda_{IMFP} \times \cos\Theta \times \ln\{1 + 1/[\beta_{ex} \times I(Si^0)/I(Si^{4+})]\}. \qquad (4.5)$$

In this equation, λ_{IMFP} for Si-2p electrons in SiO_2 is typically taken
as 2.93 nm when produced via an Al-$K\alpha$ X-ray source and β_{ex} is the
extinction coefficient. Note: β_{ex} can vary according to the collection
conditions used since this affects the collection solid cone and thereby
the percentage or electrons coming from different depths. Likewise,
λ_{IMFP} can vary, albeit to a lesser extent according to the matrix density,

etc. (see Eq. 4.1a–d). Note: Since ratios are used, the influence of surface contaminants drops out.

4.3.1.2 *Sputter Depth Profiling*

Sputtering describes the removal of surface atomic layers in a well-controlled manner as a result of the impact of energetic ions over the area of interest. This can open up regions that are initially inaccessible to XPS, that is, layers at depths well in excess of the maximum sampling depth ($3\lambda_{IMFP}$). Note: If the region of interest is less than the sampling depth, angle-resolved measurements can, in many cases, prove more effective since the sample is not damaged (a side effect of sputtering), and more accurate values of film thickness, etc., can be derived (see Section 4.3.1.1).

If sputtering is carried out in an interleaved manner with the collection of photoelectron spectra, depth profiles of the elements of interest can be obtained. Indeed, sputtering has become the most widely used method for opening up the third dimension in XPS. The most commonly used ions for sputtering in XPS are Ar^+ with energies typically in the 0.5- to 5.0-keV range. Other ions can also be used. Indeed, there is increased interest in the use of cluster ions such as C_{60}^+, C_{84}^+, $C_{24}H_{12}^+$ (coronene), or even Ar_n^+ (where n is several thousand) since these induce far less damage when sputter depth profiling of organic films, even when used at energies from 10 to 20 keV. Note: At lesser energies, deposition is often apparent. Monoatomic ions, however, presently remain the most effective for depth profiling of metal and oxide substrates.

A routine example of an XPS depth profile is shown in Figure 4.10. This was acquired from a SiO_2–TiB–Si multilayered structure over a depth of ~1 μm using 1-keV Ar^+ ions using high-energy resolution conditions (although resulting in greater scatter in the quantified plots, this provided a better sense of the speciation variations occurring as a function of depth). Also shown are spectra acquired over the B-1s, Si-2p, and Ti-2p regions. These allow chemical state variations to be followed. Note: This type of analysis can be time-consuming, that is, can extend over multiple hours.

4.3.1.2.1 *Sputter Rates*

Sputter rates pertain to the rate at which the surface is being eroded (removed) upon energetic ion impact. This can range from less than a monolayer per minute to values in excess of 10 nm/min.

In the case of monoatomic ion impact, sputtering is well understood as arising from the sequence of elastic collisions between atoms/ions making up the solid (inelastic processes, although also apparent, play

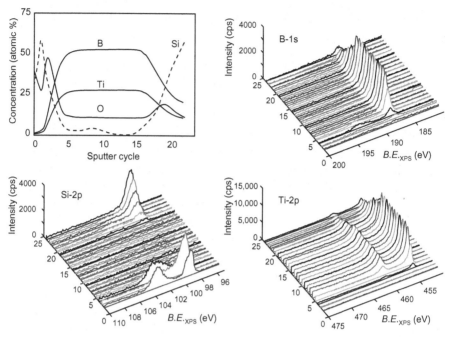

Figure 4.10. Quantified XPS sputter depth profile of a Si–TiB–Si multilayered structure (top left) along with spectra collected over the B-1s, Si-2p, and Ti 2p regions (CHA set to low-pass energy). The number on the lower left scale of the spectra pertains to the sputter cycle at which the respective spectra were collected. *Note*: Smoothing was applied to the depth profile data, and only the B, O, Si, and Ti values are shown for the sake of clarity (minor amounts of C and N were also noted). The spectra are *as recorded*.

a minor role). These ion impact-induced collision sequences are commonly referred to as *collision cascades* with the resulting sputtering described by the *linear cascade model*. Elastic scattering events resulting in sputtering are described using Newtonian mechanics.

Newtonian mechanics describes the momentum transfer occurring between two particles involved in collisions in which energy and momentum are conserved. As an example, the change in energy experienced by an ion of mass, m_1 and energy E_0 on colliding with an atom/ion of mass, m_2, that is initially at rest, can be described in the laboratory frame as (Gnaser, 1999; Sigmund, 1969, 1981)

$$E_1/E_0 = \{\cos\phi_1 \pm [(m_2/m_1)^2 - \sin^2\phi_1]^{0.5}/[1 + (m_2/m_1)]\}, \quad (4.6)$$

where ϕ_1 represents the change in trajectory experienced by the incoming ion (the scattering angle) and ϕ_2 represents the angle of trajectory

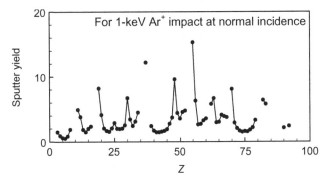

Figure 4.11. Sputter yields from elemental substrates under 1-keV Ar$^+$ impact at normal incidence (from Magee, 1981).

experienced by the collision partner with respect to the direction of the incoming ion (the recoil angle). E_1 represents the energy of the incoming ion following the collision. Since a collision cascade represents a linear combination of such events, sputtering as well as sputter rates can be derived through various simulation packages, of which SRIM (a derivative of TRIM) is the most common.

Both simulations and empirical data reveal that sputter yields arising from ion impact at a fixed energy and angle of incidence on different samples can vary by over an order of magnitude. An example of this is illustrated in Figure 4.11, in which the sputter yields from various elemental solids resulting from 1-keV Ar$^+$ impact at normal incidence are plotted versus the atomic number.

In its simplest form, sputter yields resulting from a monoenergetic ion beam incident at some constant angle can be shown to vary primarily as a function of

(a) The collisional cross section
(b) The masses of the colliding nuclei
(c) The surface binding energy

The collisional cross-section dependence is impact energy dependent, with sputter rates increasing with increasing ion energy in a continuous manner over the 0.1- to 10.0-keV range. These rates maximize between 10 and 100 keV depending on the incoming ion due to the fact that the energy transferred during the scattering of the colliding nuclei now occurs at significant depths within the solid.

At a specific energy, the collisional cross section varies with the atomic numbers of the collision partners. This dependence, relayed in

Equation 4.6, takes the form of a larger cross section for heavier elements and hence is the reason light ions are not used. As an example, the small cross section of He^+ allows this ion to travel a significant distance below the surface before interacting, hence its use in Rutherford backscattering (RBS).

The variation on atomic mass stems from the fact that the momentum transfer between collision partners increases as the masses of the two approach each other. This dependence, also revealed in Equation 4.6, will modify the form of the collision cascade. In short, if the energy transferred to the subsequent recoils remains close to the surface, an increased sputter yield will occur.

The variation with surface binding energy stems from the fact that this represents the minimum energy required to remove an atom or an ion from a solid. Such energies are usually approximated as some function of the sublimation or cohesive energy (up to 1.33 times as discussed in Gnaser, 1999). The variation stems from the fact that sublimation energies pertain to the removal of atoms from ledges and corners, while sputtering describes the removal of atoms from flat surfaces. Values are available for single-component solids. Surface binding energies for multicomponent solids are, however, more difficult to define.

Although other parameters are active, the dependence on surface binding energies can explain

(a) The lower sputter yields of oxides with respect to their base metals (oxides have larger surface binding energies than the base metal)

(b) The preferential sputtering of light elements such as oxygen and nitrogen from multicomponent solids, that is, oxides and nitrides

(c) Variations in sputter yields with crystal orientation (channeling effects, etc.)

Altering the angle of incidence also affects sputter yields and thereby sputter rates. This is understood on the basis that a greater component of the incoming ion's energy is distributed within a region closer to the surface of the solid (see Eq. 4.6) at smaller incidence angles with respect to the surface. This effect maximizes at around 20°–30° due to the introduction of

(a) The increased scattering of incoming ions

(b) The formation of surface topography

The primary impetus for decreasing the energy and incidence angle lies in the improved depth resolution that can be acquired. Sputter yields, however, also decrease with ion energy.

In the case of cluster ion impact, sputtering is less well understood. What is accepted is that the reduced sample damage noted arises from the fact that sputtering occurs primarily through lower-energy inelastic processes. (Note: On impact, the energy of a 10-keV C_{60}^+ ion will be split between the 60 atoms yielding ~167 eV per atom) Inelastic process are, however, more complex, and multiple energy deposition channels can coexist, all of which require delving more heavily into quantum mechanics.

4.3.1.2.2 Sputter-Controlled Depth Resolution Sputter-controlled depth resolution will be a function of the volume probed (sampling depth), the sample quality, and any sputter-induced damage suffered (sputter-induced diffusion, segregation, etc.).

The commonly accepted method of defining depth resolution is by measuring the depth over which a signal from some abruptly appearing layer climbs from 16% of its maximum intensity to 84%, or vice versa, as illustrated in Figure 4.12. This represents two standard deviations ($\pm 1\sigma$) of the convolution of a Gaussian function with a step function ($84 - 16 = 68\%$ as indicated in Appendix F).

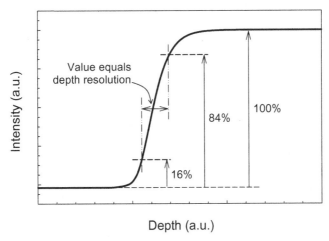

Figure 4.12. Schematic example of how the depth resolution is defined in XPS from the onset of an abruptly appearing layer (delta layer). An analogous approach can be applied as a layer of interest is abruptly terminating.

The depth resolution on well-behaved solids during sputtering under ideal conditions scales with

(a) The energy of the incoming ion. Note: Depth resolution generally improves as the inverse of the square root of the impact energy assuming no other artifacts are introduced.
(b) The angle of incidence of the incoming ion. Note: Depth resolution generally improves with a decreasing incidence angle relative to the surface, assuming no other artifacts are introduced.
(c) The temperature of the solid (this is sample specific and can have a minor effect in comparison to those listed above). Note: This approach is generally not used in controlling depth resolution.

These dependences are understood on the basis of the location and spatial extent of the collision cascade relative to the surface; that is, the smaller the volume and the closer to the surface the collision cascade is, the better the depth resolution.

Methods for reducing the ion impact-induced loss of depth resolution are, at present, limited to

(a) Altering the type of ion used
(b) Minimizing the impact energy
(c) Reducing the angle of incidence
(d) Employing sample rotation

Altering the type of ion changes the energy transferred in an elastic collision (see Eq. 4.6) and, thus, the volume over which a collision cascade extends.

Reducing the energy of the incoming ion reduces the depth of the volume influenced by the collision cascade and, thus, that suffering damage. Note: The sputtered region is always shallower than the damaged region. This explains why abruptly terminating layers always exhibit tails in their signals that extend well past the aforementioned layer.

Decreasing the angle of incidence of the incoming ion with respect to the sample surface also reduces the depth of the sputtered volume. Since a greater component of the energy is dissipated closer to the surface, an increase in sputter yield is also noted. Indeed, it is for this reason that ion guns are typically situated at 45°–30° from the sample surface. Angles smaller than this can introduce unwanted effects such as ripple topography growth, etc. (see Section 4.3.1.2.3).

As the name suggests, sample rotation employs the rotation of the sample around a point coincident with the sputtering ion beam axis during sputtering, since this enhances smoothing effects. This can be used to remove, or at least reduce, sputter-induced surface roughening and is particularly effective on polycrystalline solids.

State-of-the-art instruments are seeing depth resolution values approaching 1 nm.

4.3.1.2.3 Ion-Induced Damage

For energetic monoatomic ion impact, sputtering primarily occurs via the formation of a linear collision cascade. As described in Section 4.3.1.2.1, this is an elastic process with momentum being transferred to atoms/ions contained within the sample being analyzed. Since these ions can transfer this energy well below the sputtering front, significant damage of the sample can occur.

In the case of cluster ions (C_{60}^+, Ar_n^+ where n is several thousand, etc.), sputtering is believed to occur primarily through lower-energy inelastic processes. This is realized since, on impact, the energy of a 10-keV C_{60}^+ ion will be split between the 60 atoms yielding ~167 eV per atom. This reduced energy also results in significantly less sample damage. For this reason, the remainder of this section will concentrate on the damage realized during the linear cascades developed on monoatomic ion impact.

During a linear collision cascade, a significant amount of kinetic energy is dissipated via elastic and inelastic processes. Only a small fraction of this energy is actually removed in the form of sputtered atoms. The remainder induces a myriad of processes that culminate in the modification of both the composition (diffusion, segregation, and implantation) and the electronic structure of the solid.

Examples of energetic ion beam-induced modifications include

(a) Preferential sputtering
(b) Recoil implantation
(c) Cascade mixing
(d) Diffusion
(e) Segregation
(f) Amorphization and recrystallization
(g) Surface roughening

These processes are important in both single and multicomponent solids since these will result in the redistribution of atoms within the volume defined by the collision cascade. If significant, this can alter the compositional gradient that may have been present prior to sputtering. Furthermore, amorphization or recrystallization of either single or multicomponent solids can occur over this volume.

These aspects can be understood since the deposition of energy of greater than ~15 eV will result in the displacement of an atom bound within a lattice to some neighboring position within the lattice. If this occurs within a crystalline lattice, stable Frenkel defects (stable interstitial–vacancy pairs in close proximity) will be formed. Since sputtering represents many billions of collision cascades, all of which contain ions and recoil atoms with more than the required energy to displace lattice atoms, structural damage can result with amorphization and surface roughening being extreme examples.

Preferential sputtering is a term used to describe the preferential removal of one type of atom over another during sputtering. This occurs as a result of the difference in sputter yields experienced by different elements (see Fig. 4.11). This, in turn, stems from the preferential momentum transfer that occurs between two colliding atoms of similar masses with respect to two colliding atoms of very different masses (see Eq. 4.6). Variations in the surface binding energy will also contribute, but to a lesser degree.

An example of this is seen in the preferential removal of oxygen with a corresponding reduction of metal ions when sputtering oxides with Ar^+ ions, particularly if the metal is a high Z element. When this occurs, an altered layer is also introduced. This will result in an apparent decrease in concentration of the element experiencing preferential sputtering until steady-state sputtering conditions are reached. Thus, quantification errors will be introduced, which can, to a certain extent, be accounted for, that is, through the modification of respective sensitivity factors.

Recoil implantation describes the *anisotropic* redistribution of atoms within the solid resulting from energetic ion impact. In other words, this is an ion–atom knock-on event that transports atoms preferentially in the direction the ions are initially traveling. Since this requires excessive impact energies, extensive defect formation also occurs. As indicated by Equation 4.6, recoil implantation depends primarily on the masses of the collision partners and the impact energy. Thus, lighter elements in multicomponent solids exhibit greater susceptibility to recoil implantation at higher impact energies. Lowering the impact energy increases collisional cross sections. This enhances the scattering of heavier elements, thereby reducing the probability of knock-on events. At impact energies below 100 keV, recoil implantation becomes minimal.

Cascade mixing describes isotropic redistribution (redistribution in every direction) of atoms within solids resulting from energetic ion impact. This process dominates over recoil implantation at impact energies less than 100 keV. The isotropic nature arises from the greater number of lower-energy collisions present within a collision cascade. This also reduces the size of the affected volume and the number of defects formed. Indeed, redistribution can occur with no defect formation; that is, atoms simply swap sites over what appears as a molten region in the core of the collision cascade. This cools within 10^{-12} seconds. Cascade mixing is primarily responsible for sputter-induced loss on depth resolution (see Section 4.3.1.2.2). Indeed, the Gaussian distribution noted for layers initially exhibiting atomically sharp interfaces has been approximated using diffusion-based models (Gnaser, 1999). The diffusion coefficients derived are, however, temperature independent. Lastly, cascade mixing can be shown to act as a feeding mechanism for the preferential sputtering of atoms of different masses.

Diffusion is a thermally induced mixing process that can occur in any solid. This occurs primarily via the movement of atoms/ions through interstitial sites, vacancy sites (also called a *Schottkey defect*), dislocations (edge and screw), grain boundaries, and so on. The temperature dependence arises since the mobility of lattice defects increases with increasing temperature. Diffusion coefficients are thus a function of the number of defects present, their mobility, and lifetime.

Since sputtering introduces stable Frenkel defects within crystalline solids, diffusion is enhanced. This special form of diffusion, referred to as *radiation-enhanced diffusion*, can result in the redistribution of elements over regions well outside that of the collision cascade. In addition, this form of diffusion may or may not exhibit temperature dependence. This depends on the recombination rate of defects formed during sputtering and the annihilation rate of these defects at extended sinks (long-range defects in the form of grain boundaries, etc.). At low temperatures, the aforementioned process dominates, yielding temperature-dependent diffusion coefficients. At higher temperatures, the temperature dependence of the diffusion coefficient vanishes. At even higher temperatures, normal (thermally driven) diffusion comes into play.

Segregation is a chemically driven separation process that occurs in multicomponent solids. Gibbsian segregation is one form of segregation that describes the redistribution of elements to or from a surface in an effort to reduce the surface free energy. In other words, this describes the evolution of a surface toward its most stable state (a surface represents an abrupt termination of the long-range lattice structure). As with diffusion, Gibbsian segregation is a thermally driven process that is enhanced in the presence of defects. Recall: Defects are introduced into the lattice during sputtering.

Since sputtering removes surface layers, Gibbsian segregation is continually triggered. This is otherwise referred to as *radiation-induced segregation*. Such segregation can also result in the redistribution of elements over regions well outside the collision cascade region. Unlike radiation-enhanced diffusion, radiation-induced segregation can drive an initially homogeneous multicomponent solid to a heterogeneous state (radiation-enhanced diffusion induces the reverse process).

Amorphization/recrystallization are two opposing processes that occur as a result of the recombination or, otherwise, of Frenkel pairs. Energetic ion impact introduces Frenkel defects. If extensive enough, these can agglomerate into extended defects (dislocation loops, etc.) and can induce an initially crystalline solid to become amorphous. This is observed in brittle solids such as Si, but not in highly conductive metals. This difference stems from the fact that the distance over which Frenkel pairs recombine or, otherwise, scale with the conductivity of the solid being sputtered.

Surface roughening takes on several different forms depending on the type of solid being sputtered, its state during sputtering (crystalline or amorphous), and the sputtering conditions used. Indeed, sputtering can induce the crystallization of some solids and the amorphization of other solids.

Polycrystalline solids can also exhibit roughening due to crystal orientation sputter rate variations. An example of this is observed in the prolonged sputtering of polycrystalline zirc alloy as shown in Figure 4.13. The formation of cones, pyramids, and so on, on a particular crystal face can also occur on prolonged sputtering. These are believed to stem from the presence of foreign heavier mass atoms at the outer surface of the respective solid. If these exhibit a lower sputter yield than the surrounding surface, these will form the tip of the cone, a pyramid, and so on.

As for amorphous solids, ripple topography growth can occur under specific conditions (low impact energies and incidence angles in the case of silicon). This is believed to occur via the competing roughening and smoothing effects on/in solids that are amorphous or made to be amorphous during sputtering, for example, silicon. Roughening is thought to stem from sputter-induced surface stress, which results in surface curvature on a microscopic scale. Any variation in the incidence angle of the incoming ion will result in a variation in the sputter yields. This, in turn, will produce hillocks and valleys that propagate toward

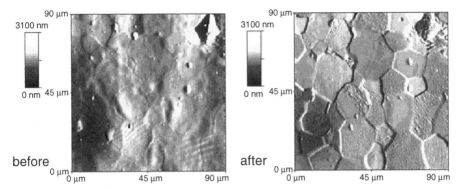

Figure 4.13. AFM images of the surface roughening resulting from extensive ion impact on polycrystalline zirc alloy. To the left is the AFM image of the polished surface prior to sputtering. To the right is the AFM image of the same area after sputtering. The location was identified through microindentation (diamond shaped crater at the top right corner of images).

the incoming ion beam like waves. Any inhomogeneities present on the surface may also play a role. Smoothing is believed to occur via diffusion apparent during sputtering, whether thermally, chemically, and/or ballistically induced.

Surface roughening is the prime cause of the loss of depth resolution often noted with increased sputtering time. For polycrystalline solids, this scales approximately with the square of the sputtering fluence. For amorphous or made-to-be amorphous solids, this takes on a steplike function.

4.4 SUMMARY

Effective XPS analysis can only be carried out with knowledge of the various factors that can influence the intensities and energies of the signals being collected. For starters, sample preparation, albeit minimal in XPS, must be carried out with care such that the deposition of contaminants is minimized.

Energy referencing is also necessary since without this, effective speciation analysis is not possible. Samples typically used for this include sputter-cleaned copper, silver, and gold metal foils with the $Cu\text{-}2p_{3/2}$, $Ag\text{-}3d_{5/2}$, $Au\text{-}4f_{7/2}$, peaks set to 932.66, 368.26, and 83.98 eV, respectively. The Ag Fermi edge from silver or the C-1s peak from adventitious carbon can also prove useful (the $B.E._{XPS}$ of the former should exist at 0 eV and the C-1s peak should lay around 285 eV).

Note: Although the C-1s $B.E._{XPS}$ is heavily sample specific, values to within 0.1 eV can be derived, which can be effective in the analysis of insulating samples.

The analysis of insulators is complicated by the fact that the energies of the signals being collected will drift if charge buildup is allowed to build up within the analyzed area. Indeed, extreme cases result in complete signal loss. Charge buildup can be controlled by irradiating the sample with low-energy electrons. Further control can be realized by immersing the sample in a magnetic field as this increases the electron current striking the analyzed area, coirradiation of the sample with low-energy ions, etc.

Damage suffered by the sample during analysis must also be minimized particularly in the case of insulators. Indeed, both X-ray and energetic electrons are ionizing forms of radiation.

The intensity of photoelectron signals is most easily understood on the basis of Spicer's three-step model. This describes how signal intensities are dependent on

(a) Photoelectron cross sections
(b) Electron interactions with the surrounding solid
(c) Instrument parameters that control the signal reaching the detector

Photoelectron cross sections are stationary state, element, and X-ray energy specific. These have been tabulated for both Al-$K\alpha$ and Mg-$K\alpha$ sources, thereby facilitating quantification.

Interaction of electrons with the surrounding solid is significant in this energy range (0–1486 eV). This typically occurs through inelastic collisions (those in which energy is lost). The IMFP represents the average distance an electron of some energy will travel between two successive inelastic collisions. Indeed, the fact that values lie between 1 and 4 nm explains the surface specificity of XPS. IMFP (λ_{IMFP}) values are best approximated using the TPP-2M relation.

Instrument parameters include such things as the angle between the source and the analyzer and the angle at which the emissions are recorded with respect to the sample surface. The former, referred to as the angular asymmetry factor, is a constant for a specific instrument. The latter, referred to as the *takeoff angle*, controls the volume from which emissions can escape the sample (also a function of the IMFP) and thus is often used to enhance, or otherwise, the surface specificity during analysis. Indeed, the sampling depth is described as the depth from which 95.7% (equivalent to a drop in intensity of e^{-3}) of all electron emissions escape the surface to be recorded. This is otherwise equal to $3.\lambda_{IMFP}\cdot\cos\Theta$ where the angle Θ is relative to the sample surface.

The analysis of elemental constituents to depths greater than the sampling depth ($3.\lambda_{IMFP}\cdot\cos\Theta$) can only be carried out by studying the cross section or through sputter depth profiling from the surface down in an interleaved manner. Sputtering describes the removal of atoms/ions comprising the sample surface through the irradiation of this surface with energetic ions (typically Ar^+ ions). Through this procedure, regions spanning up to a micron or more can be examined with relative ease. Surface layer removal in this manner can, however, result in sample damage. This stems from the fact that this form of sputtering proceeds primarily via momentum transfer. To minimize such damage, cluster ion beams are being introduced as these appear to induce sputtering through less intrusive inelastic processes.

Due to the constant nature of photoelectron cross sections and the understanding of the other parameters that control the intensities of these emissions, quantification can be a relatively straightforward procedure. Quantification errors are most typically introduced by the incorrect use of background subtraction routines and the use of ineffective λ_{IMFP} values. These errors can be compensated for by replacing the IMFP with sample-specific ALs or even by utilizing sample-specific sensitivity factors derived from reference materials of a similar composition and/or structure. Note: The base assumption in XPS is that the volume being analyzed is homogeneous. If not, the values derived will simply represent an average value over the volume probed. The background signal arises from the inelastic scattering of electrons within the solid, with electrons of nondiscrete energies escaping the solid to be recorded.

CHAPTER 5

SPECTRAL INTERPRETATION

5.1 SPECIATION

Maximizing the effectiveness of X-ray photoelectron spectroscopy (XPS) requires extracting as much information out of the raw data sets (the spectra) as possible. To do this requires

(a) An understanding of the processes active during photoelectron emission (as with any technique, the greater the level of understanding, the greater the information content that can be extracted from the raw data).

(b) Access to additional data on similar and/or related samples, whether in the form of previous or concurrent in-house studies (analysis via XPS and/or other related techniques), publications, and/or data sets. Related samples may take the form of control samples (good vs. bad), identical samples from previous studies, or systematically related samples.

Optimally, both should be used. Note: Carrying out comparative studies at the same time also provides for the most precise data sets.

X-ray Photoelectron Spectroscopy: An Introduction to Principles and Practices,
First Edition. Paul van der Heide.
© 2012 John Wiley & Sons, Inc. Published 2012 by John Wiley & Sons, Inc.

To use the example presented in Section 1.1, aluminum exists in different forms with the oxide being the most stable (oxides also form on aluminum metal surfaces under ultrahigh vacuum [UHV] conditions albeit at a slower rate). An XPS spectrum of a partially sputtered aluminum metal surface (sputtering to remove the surface oxide) at normal takeoff angles will yield overlapping Al-2p doublets within the $B.E._{XPS}$ range extending from 72.7 through 74.8 eV along with additional repeating peaks at higher $B.E._{XPS}$ values.

The individual contributions can be assigned through

(a) Angle-resolved studies (this is discussed in Section 4.2.2.2).
(b) Collection of spectra as a function of time following sputtering.
(c) Knowledge of the reasons underlying the various spectral features/trends.

Indeed, angle-resolved studies under high-energy resolution conditions will reveal an elevation in the Al-2p peak at 74.8 eV relative to that at 72.7 eV along with an increase in the oxygen content as the sampling depth is reduced. Time-dependent studies will also reveal these trends along with a more obvious suppression of the repeating structure at higher $B.E._{XPS}$. The spectral features/trends can be understood from the fact that

(a) Al-2p spin orbit splitting is below the energy resolution of instruments using standard or monochromated sources, hence the overlapping doublets.
(b) Oxidation of aluminum moves the Al-2p peaks to higher $B.E._{XPS}$, that is, induces core-level photoelectron $B.E._{XPS}$ shifts ($\Delta B.E._{XPS}$) from 72.7 to 74.8 eV.
(c) Photoelectron emissions from aluminum metal will suffer plasmon losses, hence the repeating structure noted at higher $B.E._{XPS}$ values.

Understanding these spectral features, as well as any $B.E._{XPS}$ shifts apparent, is the aim of this chapter. Indeed, this understanding is paramount in deciphering the speciation of more complex systems and is considered one of the cornerstones of XPS.

5.1.1 Photoelectron Binding Energies

Core-level photoelectron $B.E._{XPS}$ values are primarily defined by the number of protons within the nucleus and hence the value of Z of the associated atom/ion (see Fig. 2.3). As a result, the dependence of $B.E._{XPS}$

on Z can be replicated via the Sommerfeld relation, the Dirac relation (see Section 2.1.2.4), and even the $Z + 1$ approximation (see Section 5.1.1). These relations do not, however, account for the subtle yet measurable $B.E._{XPS}$ variations ($\Delta B.E._{XPS}$) exhibited by a specific atom/ion as it becomes bound within different solids or, for that matter, the excitation/relaxation of the atom/ion or its surroundings during the course of core-level photoelectron emission.

Only through complex theoretical methods involving quantum mechanics-based calculations can exact $B.E._{XPS}$ values be replicated (e.g., see Chong, 1995). This, however, is a difficult task since the bonding present and the excitations/relaxations active are all related to each other, and multiple combinations can result in the same or similar $B.E._{XPS}$ and $\Delta B.E._{XPS}$ values. In addition, no analytical technique can presently supply ground-state core-level binding energy ($B.E.$) values since all perturb the electronic structure of the element of interest in one form or another. The various contributions relating ground-state $B.E.$ values of free atoms to $B.E._{XPS}$ values of bound atoms/ions are simplistically illustrated in Figure 5.1.

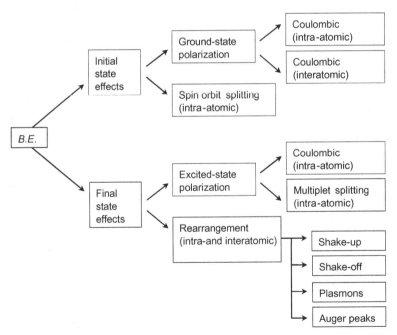

Figure 5.1. A pictorial view of all the effects that can be experienced by photoelectron emissions from a bound atom/ion with respect to the $B.E.$ value of an unperturbed free atom. As discussed in Section 4.1.3, differences in the energy reference used (E_{vac} vs. E_F) must also be accounted for.

As indicated in Figure 5.1, the $\Delta B.E._{XPS}$ values observed for a particular element present in different chemical environments can be ascribed to those arising from

(a) *Initial State Effects*: These describe the effect induced by the bonding that occurs with other atoms/ions. Note: Although only valence electrons partake in bonding, all electrons (valence and core electrons) experience the change in electron density induced. These effects are primarily responsible for the ability of XPS to derive the speciation of photoelectron emitting atoms/ions.

(b) *Final State Effects*: These describe the effect induced by the perturbation of the electronic structure resulting from photoelectron emission, particularly when core levels are involved. Since such effects also depend on the initial electronic structure (that from bonding), they too can be useful in revealing the original speciation of the photoelectron emitting atom/ion.

From Figure 5.1, it is noted that initial state effects arise from ground-state polarization (bonding) and spin orbit splitting. The former can also be subdivided into *interatomic effects* (those from neighboring atoms/ions) and *intra-atomic effects* (those from within the atom/ion). These are discussed further in Section 5.1.1.2. Likewise, final state effects can be seen to stem from photoelectron-induced *polarization* and *rearrangement* effects. Although interatomic and intra-atomic effects are again apparent, these are more closely intertwined. The same can be said for excitation and rearrangement. For example, shake-up effects can influence multiplet splitting. Final state effects are discussed further in Section 5.1.1.3.

Indeed, much confusion arose in the early days of XPS when it was assumed that $B.E._{XPS}$ values could be derived via quantum mechanics-based calculations on the assumption that Koopmans' theorem held. Koopmans' theorem states that "the negative of the eigenvalue of an occupied orbital from a Hartree Fock calculation is equal to the vertical ionization energy to the ion state formed by removal of an electron from that orbital, provided the distributions of the remaining electrons do not change" (Koopmans, 1933). This proviso is the key since this assumes that rearrangement is not in effect, which, in actuality, it is. This assumption is sometimes referred to as the *frozen orbital approach*.

To complicate matters, final state effects vary as a function of

(a) The core hole lifetime (shorter lifetimes result in stronger final state effects since these are more likely to occur within photoelectron emission timescales)

(b) The coupling that exists between electrons in different stationary states (increased coupling results in stronger final state effects), hence the enhanced final state effects noted from elements on the lower right-hand side of the periodic table (cf. Russell–Saunders vs. j–j coupling arguments)

Also of note is the fact that final state effects are always active. Their effects, however, may not necessarily be significant enough to dominate over initial state effects. This explains why $B.E._{XPS}$ values scale with the trends expected from the initial ground-state electronic configurations in some cases, and in other cases they do not. As an example, the $2p_{3/2}$ photoelectrons from titanium and the $3d_{5/2}$ photoelectrons from barium exhibit $B.E._{XPS}$ trends consistent with initial state effects; that is, the direction of the $B.E._{XPS}$ shift with oxidation state reflects the dominance of initial state intra-atomic effects as seen in titanium, or the inter-atomic effects as seen in barium (see Section 5.1.1.2.2 for further details). The Cu-2p photoelectrons from copper, on the other hand, exhibit trends that are not consistent with the initial state effects expected. The final state effects believed responsible are discussed further in Section 5.1.1.3.2.1. These $B.E._{XPS}$ shifts are plotted against the oxidation state in Figure 5.2.

Final state effects also play a part in defining the core-level photoelectron peak widths (peak widths are defined as the full width at half maxima [FWHM] of the respective peak). This is realized since the core hole lifetime (Γ) is dictated by the speed at which subsequent Auger processes occur (these result in the removal of the photoelectron-induced core hole with the excess energy removed either as an Auger

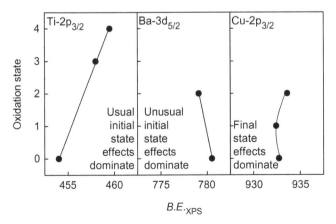

Figure 5.2. $B.E._{XPS}$ values in electronvolt from titanium, barium, and copper as a function of oxidation state. These were collected from sputter-cleaned elemental and oxide surfaces. Values are listed in Sections 6.1.2 and 6.1.4.

electron or fluorescence). Indeed, Γ has been related to the inherent core-level peak width through the uncertainty principle as

$$\Gamma = h/\text{FWHM} = 4.1 \times 10^{-15}/\text{FWHM}, \tag{5.1}$$

where the unit of Γ is second, h (Planck's constant) is electronvolt-second, and the FWHM is electronvolt (Briggs and Seah, 1990). Contributing to this will be instrumental broadening effects and photoelectron emission-induced phonons (lattice vibrations). Phonons, along with core holes, can also induce excitation of valence electrons and additional final state effects. These also influence peak shapes as discussed in Section 5.1.1.3. Indeed, core-level photoelectron peak widths extend over a large range (from a small fraction of an electron-volt to more than 1 eV) with variations even seen for the same element in different chemical environments; for example, photoelectron emissions from atoms/ions residing on the outer surface of a solid tend to exhibit broader peaks than their bulk bound counterparts.

5.1.1.1 The Z + 1 Approximation $B.E._{\text{XPS}}$ values can be roughly estimated using the $Z + 1$ *approximation* (Jolly 1977; Jolly and Hendrickson, 1970; Skinner, 1932). This procedure, also referred to as the *equivalent core approximation*, is based on the Born–Harbor cycle (Born, 1919; Harber, 1919). That used to define the Li-1s photoelectron *B.E.* is shown in Figure 5.3a with the results consistent with ab initio methods (quantum mechanics) assuming Koopmans' theorem.

The argument for lithium goes as follows: Since lithium has three electrons, two must be removed in order that one is accessed from the 1s orbital. To produce Li^{2+} requires a total energy equal to the sum of lithium's first and second ionization potential ($I_1 + I_2$). To account for the fact that the 2s electron is not actually removed in producing a Li-1s photoelectron, the 2s electron must then be placed back into the Li^{2+} ion.

The energy associated with replacing the 2s electron is, however, not equal to I_1 of Li. The $Z + 1$ approximation uses the I_2 of the $Z + 1$ element (Be) to derive this energy. This can be understood on the basis that removing an electron from a Be^+ ion (this requires an energy equal to I_2 of Be) is equivalent to inserting a 2s electron into a Li^{2+} ion. This argument is illustrated pictorially in Figure 5.3b.

Thus, the $B.E._{\text{XPS}}$ of lithium's 1s photoelectron can be approximated as $I_1 + I_2$ of Li minus I_2 of Be, that is, 81.03 ($I_1 + I_2$ of Li) − 18.21 (I_2 of Be). This can be represented as

$$B.E._{\text{XPS}} = \Sigma I_1 \cdots I_n \ (Z \text{ element}) - \Sigma I_2 \cdots I_n \ (Z + 1 \text{ element}), \tag{5.2}$$

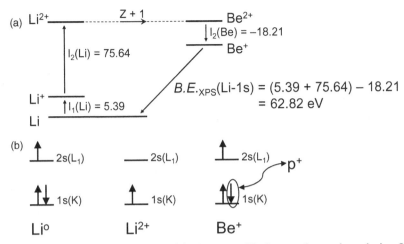

Figure 5.3. Schematic examples of (a) the Born–Harbor cycle used to derive Li-1s $B.E._{XPS}$ values, and (b) the $Z + 1$ argument for deriving the $B.E._{XPS}$ value of a Li-1s photoelectron; that is, the energy associated with placing an electron in the 2s orbit of Li^{2+} is equivalent to the I_2 of Be^+ since the extra proton (p^+) compensates for the effect of a 1s electron.

TABLE 5.1 $B.E._{XPS}$ Values Derived via the $Z + 1$ Approximation along with Empirical XPS $B.E._{XPS}$ Values for L-1s, C-1s, and Be-1s Photoelectrons in Units of Electronvolt

Element	I_n (eV)	$B.E._{Z+1}$ (eV)	$B.E._{XPS}$ (eV)
Li	75.638	62.82	64.8
C	392.077	290.49	~292.0
Ne	1195.797	843.348	870.2

where I_n refers to the value of the first electron from the level of interest. As noted in Table 5.1, the values derived for the Li-1s, C-1s, and Ne-1s levels do scale with measured values. The $B.E._{XPS}$ values obtained are, however, relative to E_{vac} since I values pertain to gas-phase atoms/ions or molecules (see Section 4.1.3).

5.1.1.2   Initial State Effects   Initial state effects describe any effect that results from the electronic structure of an atom/ion undergoing photoelectron emission that was present prior to the photoelectron emission process. These effects are split into two subgroups:

(a) Spin-induced interactions present within the photoelectron emitting atom/ion

(b) Coulombic interaction present within the photoelectron emitting atom/ion

Spin-induced interactions describe the effect a spinning charge following a nonsymmetric orbit (those with $l > 0$) has on the $B.E._{XPS}$ values of emitted photoelectrons. This is typically referred to as *spin orbit splitting*. This effect, first introduced in Section 2.1.2.5 is covered further in Section 5.1.1.2.1.

Coulombic interactions describe the influence of the charge density within the photoelectron emitting atom/ion on the $B.E._{XPS}$ values of emitted photoelectrons. This is controlled by the local chemical environment the photoelectron emitting atom/ion resides in and the electronic structure of the atom/ion itself. The former is referred to as an intra-atomic effect, whereas the latter is referred to as an interatomic effect. These effects are most commonly seen in the scaling of $B.E._{XPS}$ values with the following:

(1) Oxidation state of photoelectron emitting atom/ion
(2) Bond distance of photoelectron emitting atom with neighboring atoms/ions
(3) Madulung potentials (values only exist for ionic crystals and are only applicable to systems displaying the same structure)
(4) Electronegativity (EN) of neighboring atoms/ions (only applicable to systems of similar structure)

With the exception of spin orbit splitting, all initial state effects can be understood within the context of the *charge potential model*. This model is useful in that it describes from a classical standpoint, the interplay between intra-atomic and interatomic effects. This is discussed further in Section 5.1.1.2.2.

5.1.1.2.1 Spin Orbit Splitting All photoelectrons from orbitals described by a nonzero angular momentum quantum number ($l > 0$) will exhibit fine structure, that is, a small but observable splitting of a core level into two levels. This splitting is usually, but not always, detectable; for example, that of the 2p level of aluminum (0.4 eV) is less than the energy resolution of nonmonochromatized XPS instruments, but not synchrotron-based instruments.

Spin orbit splitting is an atomic property, and thus exhibits an element- and level-specific degeneracy and energy separation. The energy separation scales with Z as illustrated in Figure 2.4. Quantum numbers and spectroscopic notation (see Sections 2.2 and 2.3, respec-

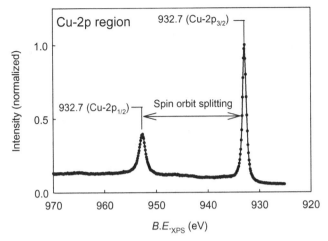

Figure 5.4. Spin orbit splitting exhibited by the Cu-2p photoelectrons from a copper metal surface. The data were collected using a concentric hemispherical analyzer (CHA) set at low-pass energy.

TABLE 5.2   Spin Orbit Splitting Factors for Electrons in the Specified Orbits (Specified Using Spectroscopic Notation), Their l and m_s Numbers, the Magnitude of J, and the Relative Peak Areas of the Spin Orbit Split Doublets

Orbital	l	m_s	J $(l + m_s)$	Area ratio $(2J + 1)$
p	1	+1/2, −1/2	3/2, 1/2	1:2
d	2	+1/2, −1/2	5/2, 3/2	2:3
f	3	+1/2, −1/2	7/2, 5/2	3:4

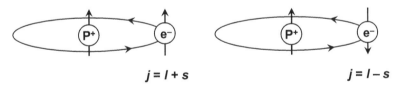

Figure 5.5. Schematic illustration of spin orbit coupling. The up arrows represent $m_s = 1/2$ and the down arrows $m_s = -1/2$. The orbit represents some nonzero orbit described by l.

tively) are used in describing the splitting as relayed in Table 5.2. An example of spin orbit splitting is shown in Figure 5.4.

Spin orbit splitting arises from the coupling of the magnetic fields produced by an electron spinning around its own axis (defined by m_s) and that produced by an electron spinning around its nucleus (defined by l) if following a nonsymmetric orbital ($l > 0$). Since m_s can have two possible values, two possible magnetic fields can be set up as illustrated in Figure 5.5.

Since spin orbit splitting only occurs for electrons from nonsymmetric orbitals ($l > 0$), the two peaks observed are described by their respective j values; that is, a spin orbit split peak is described by the level the electron emanated from followed by the value of j expressed as a subscript (spectroscopic notation). Recall: j equates to the vectorial addition of $l + s$ and $l - s$ or $l + m_s$.

The spin orbit energy separation can be approximated as (from Condon and Shortley, 1935)

$$E_{SOS} = ((2s_c + 1)/(2l + 1))K_{SOS}, \qquad (5.3)$$

where s_c is the total spin of all electrons in the orbital from which the photoelectron emanates (this takes on a value of 1/2 for all core levels since one electron is removed during photoelectron emission), and K_{SOS} is the spin orbit splitting exchange integral (this describes how strongly the two levels interact).

For a particular stationary state, this integral increases with increasing $B.E.$ This can be understood on the basis that electrons with greater $B.E.$ interact more strongly with the protons in the respective atom/ion's nuclei; that is, these are closer to the nucleus (this explains the dependence of $B.E._{XPS}$ and spin orbit splitting values for a particular stationary state and Z). Indeed, the spin orbit splitting energy should scale as $1/r^3$ (from Condon and Shortley, 1935). Since the chemical environment can also affect r, one would expect an effect on the spin orbit splitting energy. This effect will, however, be exceedingly small, that is, well below that detectable. Note: The reason why $B.E._{XPS}$ shifts are noted with the chemical environment is because the dependence of $B.E._{XPS}$ on r is much stronger; that is, this scales as $1/r$.

At this point, it is worth mentioning that the *apparent* spin orbit splitting can be influenced by multiplet splitting (this is covered in Section 5.1.1.3.1.1). This explains the apparent variations in the spin orbit splitting energies noted in the 2p spectra from cobalt and its oxides. Multiplet splitting is also responsible for the doublets observed in photoelectron emissions from s orbitals of these and related systems.

The degeneracy (population of photoelectrons taking on a specific j value) stems from the relative orientation of l with respect to s. Although requiring vector analysis for understanding, the degeneracy of electrons in p orbitals can be explained using the following highly simplified classical analogy:

Consider an aircraft flying east to west. If it experiences an updraft or down draft (force in x axis), a compensating force is required for it to remain on its initial trajectory. If it experiences a northerly or southerly wind (force in y axis) a compensating force is again required. Such a compensating force is, however, not required if the aircraft experiences an easterly or westerly wind (force in z axis).

This analogy can be used to describe the magnetic field experienced by photoelectrons in a 2p orbital since the three values of m_l can be taken to equate to the three spatial dimensions; that is, electrons in m_l states with j equal to 3/2 require an additional force to counteract the resulting force. As a result, the *B.E.* of electrons in the $2p_{3/2}$ stationary state will be less than those in the $2p_{1/2}$ state. Photoelectrons from d and f stationary states require more complex descriptions, that is, those utilizing five or seven dimensions, respectively.

The degeneracy can be approximated using Russell–Saunders (L-S) or j–j coupling arguments (from Condon and Shortley, 1935). The afore-mentioned applies to low mass elements ($Z < 30$), whereas the latter applies to high mass elements. Both fail outside their respective ranges due to coupling effects (see Section 2.1.2.5). These reveal a degeneracy that scales as

$$D_{SOS} = 2J + 1. \tag{5.4}$$

Lastly, spin orbit splitting can be thought of as an intra-atomic effect within the context of the charge potential model. The charge potential model, however, does not account for spin orbit splitting; rather, it only predicts effects of the charge density on the *B.E.* values of all peaks, with spin orbit splitting remaining constant.

5.1.1.2.2 Charge Density The charge density of electrons around the photoelectron emitting atom plays a critical role in defining photoelectron *B.E.*$_{XPS}$ values. Charge density effects are understood on the basis of Coulombic arguments; that is, a core electron *B.E.* results from the attraction it experiences to its positively charged nucleus and its repulsion to neighboring electrons. The *B.E.*$_{XPS}$ value will thus be a function of the number of electrons and protons present and the core electron density.

As an example, decreasing the electron density increases *in most cases* the respective *B.E.*$_{XPS}$ value. This can be understood from the fact that the attraction to the protons in the nucleus is distributed among fewer electrons per unit volume. This effect is illustrated in the core-level photoelectron *B.E.*$_{XPS}$ from titanium in Figure 5.2. The caveat in most cases appears since opposite trends arise when interatomic effects outweigh intra-atomic effects (these effects are discussed further in Section 5.1.1.2.2). This unusual situation explains the trends exhibited by the barium core-level photoelectron $B.E._{XPS}$ in Figure 5.2, that is, the decrease in their $B.E._{XPS}$ values with increasing oxidation state.

Since electron density can most easily be related to the EN of the neighboring atom/ion, numerous examples also exist illustrating the dependence of $B.E._{XPS}$ on EN (Siegbahn, 1970; Thomas, 1970). Some such examples are shown in Figures 5.6a,b. This scaling reflects variations in interatomic effects. Such trends are, however, limited to specific atoms/ions present in similar structures as explained further in Section 5.1.1.2.2.1.

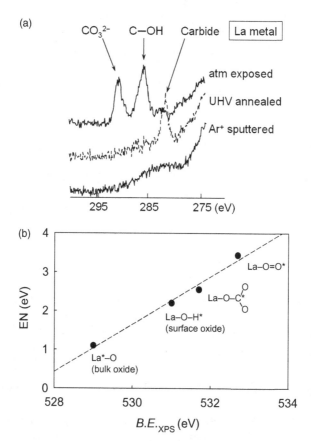

Figure 5.6. Examples of (a) photoelectron spectra collected over the C-1s region from lanthanum metal in its as-received state, after heating, and sputtering (note the scaling of the C-1s $B.E._{XPS}$ values with oxygen neighbors), and (b) O-1s $B.E._{XPS}$ values from La_2O_3 versus neighboring element EN (see asterisk) as reproduced with permission from van der Heide (2006).

Theoretical studies, even those based on Koopmans' theorem, are also able to reproduce these trends, albeit on a limited number of related systems (Gelius, 1970; Siegbahn, 1970; Shirley, 1973). Likewise for thermodynamic treatments (see Section 5.1.1.1). Indeed, thermodynamic arguments show many of the same underlying characteristics as the charge potential model (Shirley, 1972a). The charge potential model, discussed in Section 5.1.1.2.2.1, represents the most well-accepted approach for describing $B.E._{XPS}$ shifts resulting from initial state effects.

5.1.1.2.2.1 THE CHARGE POTENTIAL MODEL   The charge potential model, also referred to as the *point charge model*, is a highly simplified model that assumes atom i, the photoelectron emitting atom, is a hollow sphere of radius r_v, on which a valence charge resides. Only valence electrons are considered since only these participate in bonding. The simplicity of this model stems from the assumption that all final state effects are negligible or accounted for, and that the classical potentials at all points inside this sphere are considered equal.

This model is based on the fact that addition/removal of valence electrons will alter the coulombic attraction of all remaining electrons to the nucleus (Fadley et al., 1968). This implies the same $B.E._{XPS}$ shift ($\Delta B.E.$) will occur for all core levels within a particular atom. Indeed, this has been observed in ab initio studies (Cole et al., 2002), and in core-level $B.E._{XPS}$ of the group IA-VA elements (Wagner et al., 2003). The effect of a valence charge on the $\Delta B.E.$ exhibited by atom i as a function of the electronic structure of atom i and neighboring atoms j can then be expressed as

$$\Delta B.E._{\cdot i} \propto \left(k\Delta q_i + \Delta \sum_{i \neq j} (q_j/d_{i-j}) \right), \qquad (5.5)$$

where k is the coupling constant (defines the coulombic interaction between core and valence electrons) and d_{i-j} is a bond distance parameter (d is henceforth used to represent actual bond distances). The $k\Delta q_i$ term describes all intra-atomic effects (those induced by atom i), while the summation term describes all interatomic effects (those induced by atom j and the lattice structure). Although several forms of this relation exist, that is, k can be approximated as $1/r_v$, this form is used since it more clearly illustrates the dependence on the neighboring atom charge, q_j, and thus the Madulung potential (proportional to $q_i \cdot q_j/d_{i-j}$). Note: Since spin orbit splitting is element specific, both contributions will experience an identical $\Delta B.E.$

TABLE 5.3 The $\Delta B.E._{XPS}$ Exhibited by Group 2A Elements on Oxidation (the Level Measured Is Shown in Brackets), the Metallic Radii (r_m), the Ionic Radii (r_o), the Change in Intra-atomic Effects ($\Delta q_i/(r_m - r_o)$), the Change in Interatomic Effects ($\Delta q_j/(2r_m - \Delta d_{mi-j})$), and the $\Delta B.E$ Values Predicted from Relation 5.5

Metal	$\Delta B.E._{XPS}$ (eV)	r_{metal}	r_{oxide}	d_{i-j}	$q_i/(\Delta r_v)$	$q_j/(2r_m - d_{i-j})$	$\Delta B.E.$
Be	+2.4 (Be-1s)	1.12	0.35	1.63	−2.59	3.27	+0.68
Mg	+1.0 (Mg-2p$_{3/2}$)	1.60	0.66	2.10	−2.13	1.81	−0.32
Ca	−0.4 (Ca-2p$_{3/2}$)	1.97	0.99	2.40	−2.04	1.30	−0.74
Sr	−1.0 (Sr-3d$_{5/2}$)	2.15	1.12	2.58	−1.94	1.16	−0.78
Ba	−1.4 (Ba-3d$_{5/2}$)	2.22	1.34	2.77	−2.27	1.20	−1.07

Where Δr_v is approximated as $r_{metal} - r_{oxide}$, Δd_{i-j} as $2r_{metal} - d_{i-j}$, and Δq_i and Δq_j are −2 and 2, respectively.

Such arguments can explain the trends noted in $\Delta B.E.$ on EN and, of course, the Madulung potential. With respect to the former, increasing the EN of neighboring atoms decreases the electron density on the photoelectron emitting atom, thereby increasing $B.E.$ EN arguments are, however, limited; that is, they represent a relative elemental reactivity scale that does not always scale with r_v or d.

As an illustrative example of the utility of the charge potential model, the $\Delta B.E._{XPS}$ values noted from the group 2A elements on oxide formation are examined. The data are listed in Table 5.3. Note: The $\Delta B.E._{XPS}$ variations observed in the group 1A and 2A elements have also been explained using *partial covalency arguments* (Wertheim, 1984) and *superionicity arguments* (Barr and Liu, 1989). These assume that the partial covalency or ionicity increases on descending a group. The former can be shown to be consistent with the charge potential model in that as the covalent character of a bond increases, so does its interatomic distance (Bagus et al., 1999; Hüfner, 2003; van der Heide, 2006). These elements/substrates are considered since

(a) They exhibit significant variations in $\Delta B.E.$ on oxidation; that is, beryllium exhibits an increasing $B.E._{XPS}$, whereas barium exhibits a decreasing $B.E._{XPS}$ (values are listed in Table 5.3).

(b) The metals exhibit the same oxidation state variations (loss of two electrons).

(c) The oxides, with the exception of BeO, exhibit similar structures.

(d) The photoelectron emissions from these elements exhibit trends consistent with initial state effects (any final state effects are therefore either minimal or constant). As an example, the observed $B.E._{XPS}$ scales with the Madulung potentials of these oxides.

The first thing of note is the relative agreement in the observed $\Delta B.E._{XPS}$ trends and those predicted considering the gross simplifications used in Equation 5.5, that is, that the charge density is modeled using rudimentary geometrical arguments as opposed to the more applicable ab initio calculations.

This model also describes the variation in the $\Delta B.E._{XPS}$ observed as resulting from the interplay between intra-atomic effects and inter-atomic effects. Recall: Intra-atomic effects describe effects internal to the atom/ion while interatomic effects describe the effect of the atom's/ion's surroundings. Thus, intra-atomic effects increase the $B.E.$ when the electron density on the photoelectron atom decreases (oxidation decreases this), whereas interatomic effects increase the $B.E.$ when the electron density on the neighboring atom increases (oxidation increases this). The change listed in Table 5.3 refers to the reduction in the interatomic and intra-atomic effects, hence the change in sign.

The increase in $B.E._{XPS}$ observed in lighter group 2A elements on oxidation can thus be understood as arising from the larger change (reduction) in the interatomic effects with respect to the change in intra-atomic effects. This leaves intra-atomic effects as the dominant force. Conversely, the heavier group 2A elements exhibit decreasing $B.E._{XPS}$ since their interatomic effects decrease by a lesser amount when compared to intra-atomic effects (interatomic effects dominate).

This argument can be extended to account for the variation in O-1s $B.E._{XPS}$ values by simply reversing the signs of q_i and q_j in Equation 5.5 and accounting for the change in r_v of oxygen noted during oxide formation. The dependence of the O-1s $B.E._{XPS}$ values on $1/d$ is illustrated in Figure 5.7.

The general intra-atomic and interatomic trends, and hence the $\Delta B.E.$ exhibited by the group 2A elements, as well as that exhibited by oxygen on the formation of these oxides, can also be related to each other if one uses E_{vac} as the energy reference. This concept is expressed schematically in Figure 5.8, in which the slopes should relate to the respective element's EN. Indeed, all photoelectron emissions in which initial state effects dominate over final state effects do appear to follow these trends.

The gas phase is used as a starting point since this allows for the application of a common reference point for all atoms/ions, that is, that for which $\Delta B.E.$ from initial state effects can be assumed to be equal to zero. As a result, all $B.E.$ values, whether pertaining to the gas, liquid, or solid phase, are assumed relative to E_{vac}. This concept applies even though XPS analysis of atoms/ions present in the solid phase provides $B.E._{XPS}$ values relative to E_F since any $\Delta B.E.$ resulting from initial state effects (bonding) will not effect the position of E_{vac} (E_F can, however, be effected).

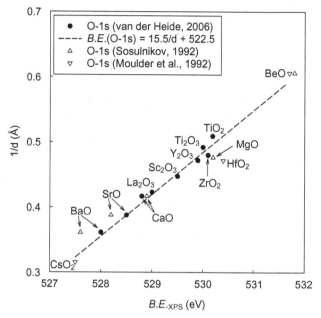

Figure 5.7. Photoelectron $B.E._{XPS}$ values as a function of inverse interatomic distance $(1/d)$ for various single-component oxides. The dashed line represents the linear regression analysis of the data. Reproduced with permission from van der Heide (2006).

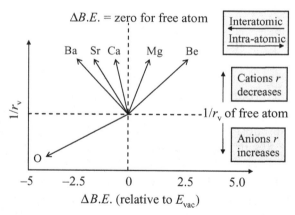

Figure 5.8. Schematic representation of the $\Delta B.E.$ expected on the oxidation of group 2A elements initially present in the gas phase (the direction of arrows points toward the oxidation of the respective element). *Note*: Atomic reference values (E_{vac}) are assumed.

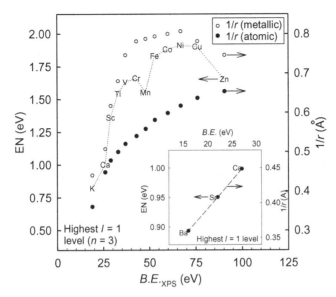

Figure 5.9. $2p_{3/2}$ $B.E._{XPS}$ values as a function of EN, inverse metallic radii, and inverse atomic radii ($1/r$) for potassium through zinc in their respective metal substrates. In the inset is plotted the Ca-$3p_{3/2}$, Sr-$4p_{3/2}$, and Ba-$5p_{3/2}$ levels from calcium, strontium and barium from the respective metal surfaces. Reproduced with permission from van der Heide (2006).

Care must however be taken when extending these arguments to other elements/systems. As an example, greater deviations are noted when plotting the $2p_{3/2}$ $B.E._{XPS}$ values of potassium though zinc from their respective metals as illustrated in Figure 5.9.

As can be seen in Figure 5.9, increased deviations from the trends expected are noted when EN or even metallic radii values are used. This can be understood on the basis that neither EN values nor the crystallographic structure has a strong influence on the internal electron density of these elements, at least not until these elements become oxidized. Crystallographic structure is mentioned since this tends to dictate the metallic radii of these elements. Note: The scaling of $B.E._{XPS}$ with EN, shown in Figure 5.6, is due to the increased range in EN values apparent and the fact that only the $B.E._{XPS}$ of closely related systems were examined. The EN trends in Figure 5.9 tend to relay the stability resulting from *Hund's rule*. In other words manganese is more stable than chromium since its 3d level is half full with all electrons of the same spin. Likewise, zinc is more stable than copper since its 3d level is full. Since this has little influence on the electron density within the core of these elements, these effects do not strongly impact the $B.E._{XPS}$

values. Atomic radii relay electron densities, hence the more consistent trends noted.

A close correlation of the $B.E._{XPS}$ values of the group 2A elements with inverse atomic radii, irrespective if atomic or metallic values are used, as well as EN, is, however, apparent as illustrated in the inset of Figure 5.9. This arises from the similar chemical reactivity noted for elements on moving down the periodic table. Elements with similar chemical reactivity are also expected to display similar final state effects.

5.1.1.3 Final State Effects
Final state effects in XPS describe any effect resulting from the perturbation to the electronic structure of an atom/ion undergoing photoelectron emission, that is, any effect (inter-atomic and intra-atomic) resulting from core hole formation. Final state effects can be subdivided into two groups, these being

(a) Core hole-induced polarization (instantaneous electrostatic and magnetic effects)
(b) Core hole-induced rearrangement (subsequent excitation and relaxation processes)

Core hole-induced polarization is an instantaneous effect that results from the removal of a spinning charge (the photoelectron) from an atom/ion. This not only results in an electrostatic effect (all electrons have unit charge) but also in a magnetic effect (spinning charges produce magnetic fields). These can influence the recorded $B.E._{XPS}$ in a number of different ways as described in Section 5.1.1.3.1.

Rearrangement effects stem from the energy transfer associated with the dissipation of the core hole produced during the photoelectron emission process. Since this and any subsequent electronic rearrangement can occur within the timescale of photoelectron emission, these can introduce new rearrangement-specific spectral features. Examples of these include

(1) Satellite peaks and photoelectron peak asymmetry
(2) Plasmon loss peaks
(3) Auger electron peaks (since these are X-ray induced, they also appear in XPS spectra, but not vice versa)

These are covered within Section 5.1.1.3.2. Note: Satellite peaks and photoelectron peak asymmetry are considered together since both stem from the same source; only the transition energies differ.

5.1.1.3.1 Core Hole-Induced Polarization Since a photoelectron has charge and spin, the core hole formed upon electron emission can be thought of as introducing opposing charge and spin. Thus, an instantaneous final state effect dependent on the electronic structure of the photoelectron emitting atom/ion is introduced.

The form of this final state effect is, however, not trivial. For example, removal of charge will increase the *B.E.* of all remaining electrons since the attraction of the remaining electrons to the nucleus is divided among fewer electrons. Increasing the *B.E.*, however, decreases the average valence r, thereby increasing the shielding of the departing photoelectron from the nucleus. This will decrease the photoelectron *B.E.*$_{XPS}$ by an amount dependent on the electronic structure and polarization suffered, all of which will be depended on the initial bonding present. To further complicate matters, no internal reference exists to which this effect can be measured.

Calculations of the *B.E.* shift expected from core hole-induced polarization reveal values of 1.5 and 2.5 eV for the Cl-2s and Na-1s photoelectron emissions, respectively (Citrin and Thomas, 1972; Hüfner, 2003). This suggests that these effects induce shifts comparable to initial state polarization.

Proof of the existence of core hole-induced polarization comes in the form of multiplet splitting. This is realized since this splitting arises from exchange interactions that occur between the odd number of electrons that remain in the core level from which electron emission took place and any unpaired valence electrons. Exchange interactions describe magnetic interactions. Since a minimum of two unpaired spinning charges is needed, only atoms/ions with unpaired valence electrons during electron emission suffer multiplet splitting.

5.1.1.3.1.1 MULTIPLET SPLITTING Multiplet splitting, also referred to as *exchange splitting*, shares similarities with spin orbit splitting in that it arises from the interaction of the magnetic fields set up by localized spinning charges. The fields responsible for multiplet splitting arise from

(a) Unpaired electrons left in the core level in which the core hole is introduced (since a core hole is required, the magnetic field associated with the remaining unpaired core electrons can equally be ascribed to the core hole itself, as is often done)

(b) Unpaired valence electrons whether initially present or introduced through relaxation (these become localized by the core hole introduced, i.e., *B.E.* values of all electrons associated with the electron emitting atom/ion increase due to the charge removal experienced)

Since valence levels play a critical role in multiplet splitting, this form of splitting is sensitive to the local chemical environment as well as any core hole-induced rearrangement effects introduced. The former stems from the fact that valence electrons are involved in bonding (see Section 2.1), while the latter explains why both paramagnetic and diamagnetic ions can exhibit multiplet splitting (Co^{2+} and Co^{3+} are two examples). Multiplet splitting can also alter the *apparent* separation induced by spin orbit splitting. As an example, the apparent spin orbit splitting of the Co-2p doublet varies from 14.97 eV noted from the metal to ~16.0 eV noted in Co_2O_3 (an example is shown in Briggs and Seah, 1990). Lastly, metals can also exhibit multiplet splitting due to core hole-induced localization of valence electrons. Note: Valence electrons must be localized in order to induce multiplet splitting.

Examples of multiplet splitting are shown in the 3s photoelectron spectra of Cr_2O_3, Fe_2O_3, and CuO in Figure 5.10.

At the most fundamental level, multiplet splitting arises from the fact that an electron can have one of two spins (m_s can equal +1/2 or −1/2), which in turn generates a specific magnetic field. If two unpaired electrons exist (valence and core), the respective magnetic fields can be aligned or opposed to each other. This results in a splitting in energy of

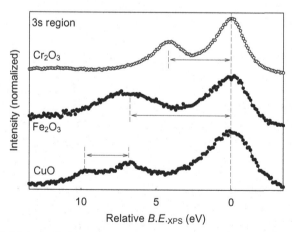

Figure 5.10. Multiplet splitting over the transition metal 3s regions from Cr_2O_3, Fe_2O_3, and CuO (CHA set to low-pass energy). The splitting is represented by the arrows. For comparative purposes, the main photoelectron $B.E._{XPS}$ is set to zero. *Note:* The splitting of the Cu-3s spectrum from CuO occurs on the 3s satellite, not the main peak as in Cr_2O_3 and Fe_2O_3.

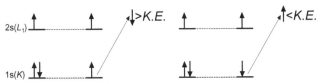

Figure 5.11. Schematic illustration of muliplet splitting in the 1s photoelectron emissions from lithium for 2s electrons of the same m_s. Combining these with those of the opposite m_s (2s electron) results in four possible orientations, all of which result in two different Li-1s energies.

the associated orbitals and thus the $B.E._{XPS}$ of the core level from which the initial electron emanated from (that responsible for the core hole in the first place). Unlike spin orbit splitting, multiplet splitting can affect the $B.E._{XPS}$ of any electron from any orbital, even s orbitals.

An example of this interaction is illustrated for Li-1s photoelectrons in Figure 5.11. The one electron left in the 1s level can have m_s equal to +1/2 or −1/2. This will set up a magnetic field that will interact with the magnetic field induced by the lone valence electron, that is, that in the 2s level, which can also have m_s equal to +1/2 or −1/2. These magnetic fields will split the energy of the 1s orbital, in this case, into two discrete values. Similarities between this and electron spin resonance (ESR) are thus apparent. Note: In ESR, an externally applied magnetic field takes the place of that induced by the core hole.

The multiplet splitting energy separation (E_{mult}) on a the core-level photoelectron emission can be shown, via vectorial analysis, to scale as

$$E_{mult} = ((2(S_v + s_c) + 1)/(2l + 1))K_{mult}, \qquad (5.6)$$

where K_{mult} is the multiplet splitting exchange integral. This integral increases as the interacting core electron $B.E.$ decreases, or in other words, as the unpaired core and valence electrons move toward each other. Note: The inclusion of the core-level spin (s_c) in Equation 5.6 is a recently suggested correction (van der Heide, 2008) to account for the underestimation of E_{mult} previously noted (previous relations assumed the form originally developed to describe the effect of a valence hole [van Vleck, 1934], not a core hole).

Since the extent of splitting depends on the exchange integral, this and the value of E_{mult} can be related to the $B.E._{XPS}$ of the level affected by multiplet splitting. The value of E_{mult} for the 3s level from the first row transition metal ions appears to scale as (van der Heide, 2008)

$$E_{mult} = ((2(S_v + s_c) + 1)/(0.6(B.E._{XPS})^{1/9}). \qquad (5.7)$$

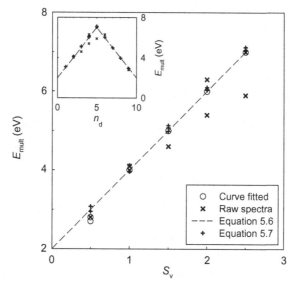

Figure 5.12. Multiplet splitting energy (E_{mult}) versus S_v. In the inset is shown E_{mult} versus the number of electrons in the transition metal ion 3d level (n_d). The dashed lines represent Equation 5.6 with the sum of s_c and S_v represented by S. Also shown is the relation between E_{mult} and the $B.E._{XPS}$ as relayed through Equation 5.7 (figure reproduced with permission from van der Heide, 2008).

As can be seen, the splitting pattern is a function of the core level l; that is, photoelectrons from s orbitals result in doublets, p orbitals in triplets, d orbitals in quintuplets, and so forth. Similar trends can be derived (empirically or otherwise) for other levels and other transition metal ions.

The extent of the splitting apparent in the 3s levels of the first row transition metal ions is summarized in Figure 5.12 and in its inset. This reveals the scaling with the total 3d level spin (S_v) of the photoelectron emitting atom/ion as would be expected from the above arguments. In addition, the first row transition metal ions with less than half-filled 3d orbitals exhibit a splitting of the main photoelectron peak, while those with more than half-filled 3d orbitals exhibit a splitting of the satellite peak (satellites are covered in Section 5.1.1.3.2.1). Note: When spin orbit splitting is in effect, only the spin orbit peak with the higher j value suffers multiplet splitting.

Multiplet splitting degeneracy can be expressed via the same relation used to describe spin orbit splitting degeneracy (Eq. 5.4) if all spins are effectively accounted for. Although normally represented using J (van Vleck, 1934), this is represented for $l = 0$ levels using S_v and s_c (for consistency's sake) as

$$D_{\text{mult}} = 2(S_v \pm s_c) + 1 \qquad (5.8a)$$

or

$$D_{\text{mult}} = 2J + 1. \qquad (5.8b)$$

The interaction of an s core level with a single valence electron will therefore result in a 3:1 degeneracy. Likewise, two valence electrons of the same spin will yield a 2:1 degeneracy, and so forth.

5.1.1.3.2 Rearrangement Photoelectron emission introduces a core hole into the electronic structure of the photoelectron emitting atom/ ion. Core holes are, however, an extreme form of excitation that dissipates readily, that is, within 10^{-14} seconds, with the energy carried away through the emission of additional electrons (Auger emission) and/or photons (fluorescence) as discussed in Section 5.1.1.3.2.4.

Core hole formation can also lead to excitation of valence electrons associated with the photoelectron emitting atom/ion. These can then relax back into their original state or some other state; which state these relax into depends on the electronic structure of the system and hence the initial bonding present. Multiplet splitting can reveal which states are involved (see Section 5.1.1.3.1). Multiplet splitting also reveals that the excitation/relaxation of valence electrons can proceed on a timescale that is as fast, or faster, than the process leading to the dissipation of core holes, that is, that responsible for Auger and/or photon emission.

Rearrangement effects influence the $K.E._{\text{XPS}}$ of photoelectrons as they depart the photoelectron emitting atom/ion. This can be understood using the following arguments. If a valence electron were to be excited within the timescale of photoelectron emission, it will reduce the $K.E._{\text{XPS}}$ of the outgoing photoelectron by that same amount. If this excited valence electron were then to return back to its ground state or some other lower-energy state (interatomic and/or intra-atomic relaxation channels may be accessible), the de-excitation energy may then be transferred back to the departing photoelectron, thereby further influencing the measured $K.E._{\text{XPS}}$ and thus the resultant $B.E._{\text{XPS}}$ value derived.

Since more than one final state can arise, more than one photoelectron peak can be observed from a specific level of a specific element. When this occurs, the photoelectron peak appearing at a lower $B.E._{\text{XPS}}$ is commonly termed the *no loss peak* or the *main peak*. All other peaks/ contributions (those at higher $B.E._{\text{XPS}}$) are generally referred to by the process responsible. These are commonly grouped into the following:

(1) Shake-up processes (excitation of valence band electrons)
(2) Shake-off processes (removal of valence band electrons)
(3) Plasmon generation (excitation of conduction band electrons)

These, along with Auger electron emission, are discussed in Sections 5.1.1.3.2.1 through 5.1.1.3.2.4.

5.1.1.3.2.1 SHAKE-UP SATELLITES Shake-up processes result from transitions of valence electrons to discrete and/or nondiscrete levels located close to E_F. These processes result in the introduction of either or both

(a) Satellites that can be of measurable intensity
(b) Asymmetries on the main photoelectron peak

Since such processes reduce the $K.E._{XPS}$ of departing photoelectrons, both satellites and peak asymmetry are noted at higher $B.E._{XPS}$ with respect to the associated main photoelectron peak. The only difference lies in the energy separation; that is, the latter occurs if the energy between the main and satellite peaks falls below that detectable under the analytical conditions used. In the former, the energy separation is sufficient to distinguish the satellite from the main peak.

Shake-up features can be useful in providing information on an element's speciation. An example of this is illustrated in Figure 5.13,

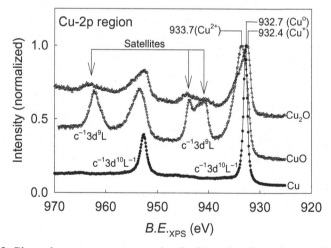

Figure 5.13. Photoelectron spectra over the Cu-2p region from the metal and oxides (Cu_2O and CuO) of copper collected under high-energy resolution conditions (CHA set to low-pass energy).

in which spectra from CuO exhibits appreciable shake-up satellites, whereas spectra from copper metal and Cu_2O do not. Note: The $B.E._{XPS}$ of the main Cu-2p photoelectron peaks also do not follow the trends expected from initial state effects, that is, do not scale with oxidation state (see also Fig. 5.2).

The shake-up features noted over the Cu-2p region from Cu^{2+} in CuO have been described by the Kotani–Toyozawa model (Kotani and Toyozawa, 1974). This model stipulates that these satellites arise from electron transitions from filled ligand states to the unfilled states situated close to E_F, that is, from the O-2p valence band to the 4sp conduction band.

These satellites are referred to as the $c^{-1}3d^9L$ peaks where the L term represents the ligand electron and the c^{-1} term represents the core hole introduced (the notation used is also discussed in Section 2.1.2.4). Note: The c^{-1} term is sometimes not included; that is, it is assumed.

The main Cu-2p photoelectron peak then arises from a subsequent final state formed when the excited ligand electron drops into the Cu-3d valence band. These peaks are thus referred to as the $c^{-1}3d^{10}L^{-1}$ peaks, where the L^{-1} term signifies the transfer of an electron from the ligand into the Cu-3d band. This is described as a *charge transfer* process.

These perturbations have been simplistically described as a $c^{-1}d^9L \rightarrow c^{-1}d^{10}L^{-1}$ transition. Since electron transfer occurs, one of the photoelectron peaks from a specific Cu core level suffers multiplet splitting (the $c^{-1}d^9L$ state), whereas the other does not (the $c^{-1}d^{10}L^{-1}$ state). Recall: Multiplet splitting only affects the higher j peaks in $l > 0$ levels (see Section 5.1.1.3.1).

The underlying reason why oxides such as CuO are believed to suffer charge transfer can be traced back to their optical properties, or more precisely their bandgaps (E_g). This is asserted on the basis that the transitions believed responsible can be represented as $d^9L \rightarrow d^{10}L^{-1}$ (Imada et al., 1998). The energy between the d^9L and $d^{10}L^{-1}$ states is generally represented by the theoretical parameter Δ (~4 eV for CuO from Zannan et al., 1985). The difference between E_g and Δ arises since Δ does not account for hybridization or transitions into the 4sp band. Optical methods are used in defining E_g since core holes and their effects are not introduced, and these oxides have E_g values within the ultraviolet range (<6.7 eV).

Note: The 4sp band is often left out of such descriptions, as well as many ab initio calculations used to model these *many body* effects (many electron), since the simplified two-level approximation ($d^9L \rightarrow d^{10}L^{-1}$) appears effective. In other words, an electron in the delocalized 4sp band has essentially the same effect as if it were to remain in the

O-2p band (neither screen the Cu core levels nor induce multipet splitting). In actuality, the transfer appears to proceed via the 4sp band (Imada et al., 1998).

Not discussed thus far is the effect of the *on-site coulombic repulsion* of electrons within a specific band (Mott, 1949; Hubbard, 1963). This effect splits the 3d band into what is referred to as a lower Hubbard band (LHB) and an upper Hubbard band (UHB), with electrons in the LHB becoming more localized, while those in the UHB becoming more delocalized. The latter can also rise above E_F.

The energy difference between these bands is generally represented by the theoretical parameter U_{dd} (~7 eV in the case of CuO). Since this value decreases slightly with the number of electrons in the 3d band while the value of Δ increases, transitions between these states are believed to become the dominant factor in defining E_g in the early transition metal oxides (Imada et al., 1998; Zannan et al., 1985).

When optically induced, this transition can be represented as $d_i^n + d_j^n \rightarrow d_i^{n-1} + d_j^{n+1}$, where the subscripts i and j refer to adjacent metal ions and n refers to the initial 3d population. Core holes will modify this transition to $c^{-1}d_i^n + c^{-1}d_j^n \rightarrow c^{-1}d_i^{n-1} + c^{-1}d_j^{n+1}$, with the $c^{-1}d_i^n$ and $c^{-1}d_i^{n-1}$ configurations expected to exhibit different $B.E._{XPS}$ values. Note: It can be argued that these transitions proceed via the 4sp band since this would delocalize these electrons before being trapped on an adjacent metal ion site.

Since Mott–Hubbard transitions appear to define the value of E_g in the early transition metal oxides and charge transfer defines the value of E_g in the late transition metal oxides, the former is commonly referred to as Mott–Hubbard compounds, while the latter is referred to as charge transfer compounds (Zannan et al., 1985). Clear examples of the former include Ti_2O_3, V_2O_3, VO_2 (below 338 K), and Cr_2O_3. The latter include NiO and CuO. Compounds with similar Δ and U_{dd} values exhibit optical properties that lie between these extremes. These are referred to as *intermediate compounds*, with Fe_2O_3 and MnO being two accepted examples (Hüfner, 2003; Imada et al., 1998; Zannan et al., 1985).

Schematic illustrations of the transitions believed to be induced on core hole formation in V_2O_3 and CuO are shown in Figure 5.14a,b. Of note is the fact that 3d electrons are lost from photoelectron emitting ions suffering Mott–Hubbard transitions, while 3d electrons are gained by photoelectron emitting ions suffering charge transfer. Also, U_{cd} reduces the energy needed for charge transfer (this does not appear to effect Mott–Hubbard transitions as strongly).

Valence photoelectrons do not suffer these extensive rearrangement effects since core holes are not introduced. This explains why, for example,

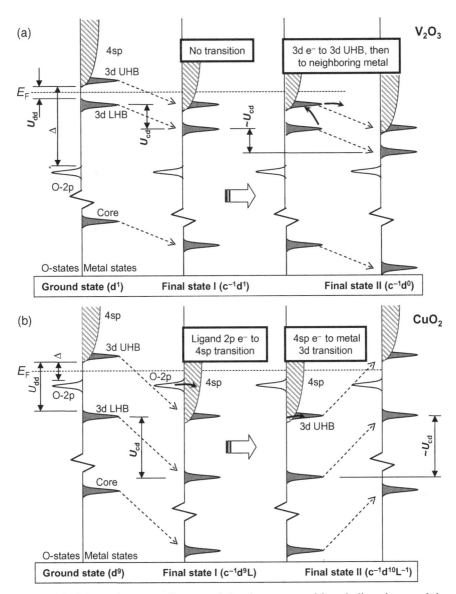

Figure 5.14. Schematic energy diagram of the electron transitions believed to result in the final states following photoelectron emission from (a) V in V_2O_3 initially in its ground state and (b) Cu in CuO initially in its ground state. Ligand levels are portrayed as white (2p), while the metal levels are portrayed as gray (3d) and hatched (4s). The transitions, depicted by the curved arrows, proceed to fill levels below E_F/vacate levels above E_F (Aufbau principle). All levels are arbitrarily referenced to a common E_F (that of the surrounding solid), and all definitions are described in the text (figure reproduced with permission from van der Heide, 2008). The position in the levels of the photoelectron emitting atom/ion moves as a result of the variation in electron densities occurring.

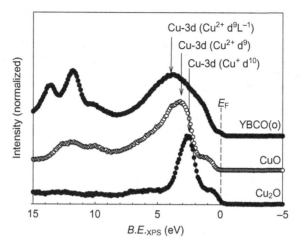

Figure 5.15. Valence band spectra from Cu present within Cu_2O, CuO, and YBCO(o) collected under the same conditions as spectra in Figure 5.13 (Al $K\alpha$ source used). YBCO(o) refers to the orthorhombic phase of yttrium barium copper oxide.

the Cu-3d $B.E._{XPS}$ values appear to scale more effectively with the oxidation state. An example of this is illustrated in Figure 5.15, in which the the Cu-3d spectra from Cu_2O, CuO, and yttrium barium copper oxide (YBCO). Note this scaling is not seen for the Cu-2p $B.E._{XPS}$ (see Fig. 5.13).

For this reason, UPS is effective in examining density of states (DOS) and even the superconducting gap (see Hüfner, 2003). The latter, however, requires highly specialized instrumentation.

In the case of systems in which bonding is primarily covalent in nature (bonding resulting from electron sharing), a different approach is generally taken to describe the satellites often noted. This employs the use of bonding and antibonding states (σ and π bonding states and σ^* and π^* antibonding states) with satellites described as arising from transitions between the bonding and antibonding states. An example of this is shown in Figure 5.16, in which the broad satellite structure noted ~6 eV above the main C-1s peak is typically described as resulting from π to π^* transitions.

Also apparent in Figure 5.16 is the asymmetry of the main C-1s peak. This arises from the energy associated with the transitions of valence electrons into states that are within the energy not resolvable by the instrument (this can be a sizable fraction of an eV in non-synchrotron-based instruments). These transitions maybe electronic, vibrational, or even vibronic (rotational states are not considered since these are not resolvable in XPS). In the case of the C-1s photoelectron peak from

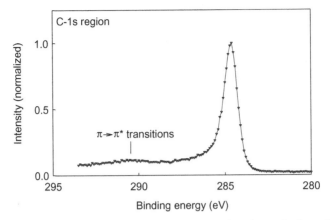

Figure 5.16. Photoelectron spectra over the C-1s region from single-walled carbon nanotubes (CHA set to low-pass energy). Of note is the shake-up peak at ~6 eV above the main C-1s peak and the asymmetry introduced into the high $B.E._{xps}$ side of the main C-1s peak.

covalently bound carbon, the asymmetry is typically ascribed to vibrational excitation.

Note: Core-level photoelectron peaks not suffering rearrangement effects should be symmetric in shape with the width (peak FWHM) defined by the Auger decay rate (uncertainty principle as indicated in Eq. 5.1), photoelectron-induced phonons, and the instrument. Phonons represent the energy associated with lattice vibrations. In XPS, phonons arise from the photoelectron-induced ionization process (the associated ionic radii are smaller than the atomic radii), as well as the photoelectron recoil effect (from conservation of momentum). The latter, however, can be considered negligible when using conventional XPS sources due to the large electron to atom/ion mass difference.

5.1.1.3.2.2 SHAKE-OFF SATELLITES   Shake-off processes result when electrons are excited from valence levels into unbound continuum states located above E_{vac}. As a result, these contribute to the broad background at higher $B.E._{xps}$ values with respect to the main photoelectron peak (discrete shake-off peaks are rarely observed). The difference between shake-off and shake-up processes lies in the fact that electrons depart the system in the former but not in the latter.

5.1.1.3.2.3 PLASMON LOSS FEATURES   Substrates that exhibit a high density of free electrons around E_F during core level electron emission

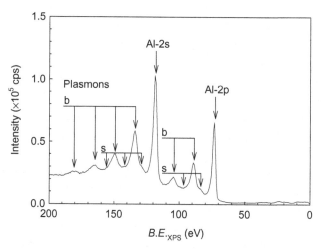

Figure 5.17. The plasmon loss structure noted over the Al-2s and Al-2p regions from a sputter-cleaned aluminum metal (CHA set to low-pass energy). The surface and bulk plasmon losses are indicated as s and b, respectively.

are susceptible to plasmon formation. Extreme examples of this are seen in the spectra of Mg, Al, and Si when present in their elemental solids. Insulators such as TiO_2 can also exhibit plasmon formation since core-hole formation can induce the transfer of valence electrons into the conduction band (see Section 5.1.1.3.2.1).

An example of how plasmon loss features affect photoelectron spectra is shown in Figure 5.17. As can be seen, both surface plasmons and bulk plasmons are noted at repeating $B.E._{XPS}$ values (surface plasmons at intervals of 11.2 eV and bulk plasmons at intervals of 15.8 eV) greater than the associated main photoelectron peaks (Al-2p and Al-2s, in this case).

Plasmons are collective oscillations in the sea of electrons, that is, all the electrons around E_F. These are formed as conduction band electrons move to account for a sudden change in charge density occurring on photoelectron production. In doing so, these electrons can overshoot their mark, thereby setting up *collective oscillations* in the conduction band. The sudden change in charge density can arise from either the passage of fast charged particles (electrons) or the formation of a core-hole (that resulting from photoelectron emission). Plasmons resulting from the former are referred to as *extrinsic plasmons*, while those resulting from the latter are referred to as *intrinsic plasmons*.

When free electrons oscillate, they prefer to do so at some specific frequency and energy. These are termed the *free electron plasmon fre-*

TABLE 5.4   Bulk Plasmon Energies (from Eq. 4.1b) for Various Elements within Their Elemental Solids along with N_v, ρ, and M Values

Solid	E_{pl} (eV)	N_v	ρ (g · cm³)	M (g/mol)
C	24.83	4	2.23	12.000
Na	5.92	1	0.97	22.990
Mg	10.87	2	1.74	24.350
Al	15.78	3	2.70	26.982
Si	16.59	4	2.33	28.086
Sr	7.02	2	2.60	87.62

quency and the *free electon plasmon energy.* Assuming no interband transitions exist close to the plasmon energy (see below), the frequency and energy associated with collective oscillations within the bulk (3D plasmons) and surface (2D plasmons) of a solid become a function of the density of free electrons in the respective metal. Bulk plasmon energies are defined (from Ritchie 1957) in Equation 4.1b. Surface plasmon energies are defined as the bulk value divided by $\sqrt{2}$.

Listed in Table 5.4 are $E_{pl(bulk)}$ as derived from Equation 4.1b along with the parameters N_v, ρ, and M used in solving this relation for core level photoelectron emissions for various elements within their respective elemental solids.

If interband transition energies approach $E_{pl(bulk)}$, plasmons will dissipate through the promotion of an electron between the respective bands. In the case of Ag, there exist levels within the valence region that have an energy separation close to 4 eV (this represents the transition energy for electrons in the $4d_{3/2}$ and $4d_{5/2}$ levels to E_F as defined by the 4p-5s band). This transition has the effect of reducing the values of $E_{pl(bulk)}$ and $E_{pl(surface)}$ from those specified above, to values equal to the interband transition energy. Silver thus exhibits surface and bulk plasmon loss peaks close to 4 eV apart (Pollak et al., 1974), even though Equation 4.1b specifies $E_{pl(bulk)}$ values well in excess of this. The same is noted for the surface plasmon loss peaks.

5.1.1.3.2.4   AUGER PEAKS   Photoelectron emission is always accompanied by Auger electron emission since this, along with fluorescence, represents the primary route for the dissipation of energy introduced by core hole production. Auger electron emission is observed for all elements containing three or more electrons (Li–U) with the emission probability being a function of Z and the stationary states involved. Since both Auger electron emission (Auger, 1925) and fluorescence (Röntgen, 1896) were described by physicists, X-ray notation is used in describing the respective transitions (X-ray and spectroscopic notation are covered in Section 2.1.2.3).

Auger electron emission is a three-electron process. The first electron represents the photoelectron. The second and third electrons remove the energy associated with the core hole; that is, the second electron drops down from some level of lesser *B.E.* to fill the core hole and the third electron removes the energy associated with this transition. This is illustrated in Figure 1.1a for the KL_2L_3 Auger electron emissions from oxygen (the KL_2L_3 designation reveals a K-level hole is filled by an electron from the L_2 level with the energy difference carried away through the emission of an electron from the L_3 level).

Other transitions are also possible. Indeed, three distinct Auger peaks are observed for oxygen in both XPS and Auger electron spectroscopy (AES). These are otherwise referred to as the KL_1L_1, $KL_1L_{2,3}$, and $KL_{2,3}L_{2,3}$ peaks resulting from the KL_1L_1, KL_1L_2, KL_1L_3, KL_2L_3, and KL_3L_2 transitions. Only three peaks are observed since the KL_1L_2 and KL_1L_3 as well as the KL_2L_3 and KL_3L_2 transitions result in electron emissions of similar energies (hence the aforementioned $KL_1L_{2,3}$ and $KL_{2,3}L_{2,3}$ peak designations). The fact that all of these transitions are observed reveals that the dipole selection rules adhered to in fluorescence are not strictly obeyed in Auger electron emission. This likely arises from the greater perturbation suffered as a result of additional core holes being present.

The dipole selection rules, which specify transition probabilities, are defined as

$$\Delta l = \pm 1, \Delta m_1 = 0, \pm 1.$$

As an example, these would *forbid* KL_1L_1 transitions. Note: *Forbid* in this context means to have low probability. Indeed, KL_1L_1 peaks are noted in XPS spectra although displaying weaker intensities than the oxygen $KL_1L_{2,3}$ and $KL_{2,3}L_{2,3}$ peaks.

The measured energy of the Auger electron equates to the energy difference between the two levels accessed by the second electron minus ϕ. As described in Section 4.1.3, ϕ is taken as that of the instrument since this is typically less than that of the sample. Since all electron emissions in XPS are plotted on a *B.E.*$_{XPS}$ scale, any Auger electron emissions recorded will also exhibit some apparent *B.E.*$_{XPS}$. This, however, is a fictitious value since the Auger electron's true kinetic energy (*K.E.*) represents the difference between the two levels accessed minus ϕ. Thus, if a different energy photon source is used, the Auger electrons will appear at a different *B.E.*$_{XPS}$ (this can be used to identify

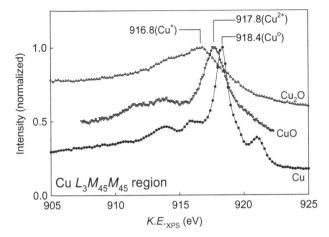

Figure 5.18. X-ray-induced $L_3M_{4,5}M_{4,5}$ Auger peaks from metal and oxides (Cu$_2$O and CuO) or copper (CHA set to low-pass energy). *Note*: The energy scale used is converted to $K.E._{XPS}$.

Auger electrons vs. photoelectron peaks). To be technically correct, Auger electron emissions should be plotted on a $K.E._{XPS}$ scale. This, however, is not feasible in XPS (both energy scales cannot be used at once). The $B.E._{XPS}$ scale is therefore used since Auger electrons are of lesser importance relative to photoelectron peaks in XPS.

X-ray-induced Auger electron emissions can provide information on speciation. An example of this is shown in Figure 5.18, in which the Auger electron emissions from the metal and oxides (Cu$_2$O and CuO) or copper are shown. The electron transitions responsible are between the Cu-$M_{4,5}$ (Cu-3d$_{5/2}$, 3d$_{3/2}$) and Cu-L_3 levels (Cu-2p$_{3/2}$).

As can be seen in Figure 5.18, variations in both their peak positions and structure are noted. The Cu $L_3M_{4,5}M_{4,5}$ region is of particular interest in the analysis of the oxidation state of Cu since, as discussed in Section 5.1.1.3.2.1, the final state effects active on the Cu-2p photoelectron emissions obscure this information. Note: Initial and final state effects are also apparent in Auger electron emission. These, however, differ from those active on photoelectron emissions since two core holes are now present. This is discussed further in Section 5.1.1.4.

5.1.1.4 The Auger Parameter
Another means of extracting information on speciation, as well as to examine the final state effects active on photoelectron emission, lies in the analysis of *Auger parameters* (e.g., see Moretti, 1998). The original form of this was taken to represent the sum of the Auger electron $K.E._{XPS}$ plus the photoelectron $K.E._{XPS}$ (Wagner, 1972). To avoid the generation of negative values, the photon

energy was then added (Gaarenstroom and Winograd, 1977; Wagner et al., 1979). This form, then referred to as the *modified Auger parameter*, represents the sum of the Auger electron $K.E._{XPS}$ and the photoelectron $B.E._{XPS}$.

Since this definition is now used with the *modified* term dropped, the Auger parameter term will be applied henceforth to represent the modified form. In other words, the Auger parameter for the Cu^{2+} emissions from CuO will be taken to be equal to the sum of the Cu $L_3M_{4,5}M_{4,5}$ Auger electron $K.E._{XPS}$ (917.8 eV) plus the Cu-$2p_{3/2}$ $B.E._{XPS}$ (933.7 eV) or 1851.5 eV. This type of approach can be represented graphically by plotting the Auger electron $K.E._{XPS}$ against the photoelectron $B.E._{XPS}$, as demonstrated in Figure 5.19.

These are typically referred to as *Wagner plots*. Of note is the fact that the data can fall along two distinct lines with one exhibiting a slope of unity and the other exhibiting a slope of 3. Intermediate values can also exist. The former will only occur if initial state effects were to dominate, whereas the latter occurs if final state effects were to

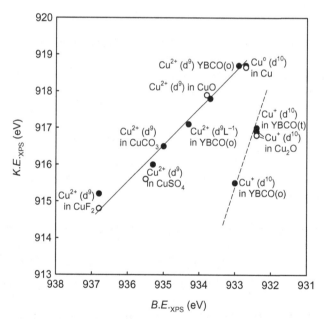

Figure 5.19. Cu-$2p_{3/2}$ photoelectron $B.E._{XPS}$ from various related oxides versus their respective Cu $L_3M_{4,5}M_{4,5}$ Auger electron $K.E._{XPS}$. YBCO(o) and YBCO(t) refer to the orthorhombic and tetragonal phases of YBCO, respectively. The data used in constructing this plot are listed in Section 6.1.4. These plots are otherwise referred to as *Wagner plots*.

dominate. Note: Only one core hole is produced during photoelectron emission, whereas two core holes are produced during Auger emission.

On the basis of these simplistic arguments, initial state effects can be inferred to be responsible for any shifts if the difference in the shifts remains independent of the number of core holes present (slope of unity). Likewise, final state effects can be inferred to dominate if larger shifts are noted with the number of core holes.

Lastly, there are situations in which Auger parameters appear difficult to obtain. One example lies in the analysis of the Si-2p emissions since the related Auger signal lies at an apparent negative $B.E._{XPS}$ value when using either Mg-$K\alpha$ or Al-$K\alpha$ sources. This issue can be overcome by collecting data over the appropriate $B.E._{XPS}$ range (may require deluding the instrument's software, that is, adjusting the value of E_{ph} from that actually used in data collection) or by using higher-energy photon sources, that is, Ag or Cr sources.

5.1.1.5 Curve Fitting   Curve fitting is a method often used in XPS to separate overlapping contributions that may be present within a prespecified $B.E._{XPS}$ region. This may need to be used if there exists

(a) Overlapping core-level peaks from different elements. This problem tends to manifest itself in spectra from the lighter elements since these exhibit fewer useful peaks.

(b) Overlapping core-level peaks due to the same element present in multiple environments

It must, however, be emphasized that curve fitting is *not* a true deconvolution procedure. This is made evident in the fact that a number of different outcomes are possible.

To obtain the curve fit that most closely describes the system under study, the physical significance of the outcome should be ensured. This can only be accomplished through knowledge of the composition and chemical state of the system and how this is modified by various instrument conditions, for example, takeoff angle and heat treatments. Following the steps listed below can ensure the above criteria are met:

First, understand and apply the appropriate spin orbit splitting energy separation and degeneracy if present (spin orbit splitting is discussed in Sections 2.1.2.5 and 5.1.1.2.1).

Second, understand and account for any

   (a) Multiplet splitting effects present

   (b) Contributions introducing an asymmetry into the photoelectron peaks of interest

At this point, it should be emphasized that core-level photoelectron peaks not suffering rearrangement effects or overlaps with other peaks should be symmetric in shape around their mean $B.E._{XPS}$ value. Indeed, these should exhibit a predominantly Gaussian form with Lorentzian tails (70%–90% of the former). The Gaussian form stems from the effects of the phonons introduced on photoelectron emission and instrument broadening effects (analysis condition specific). The Lorentzian form, on the other hand, arises from the uncertainty principle associated with the core hole lifetime (see Eq. 5.1).

Asymmetry can be introduced into the photoelectron peak shape if

   (a) There exist shake-up processes (electronic, vibrational, or vibronic transitions) whose energies are less than the energy resolution of the instrument used. Note that the asymmetry is introduced on the high $B.E._{XPS}$ side of the main peak with an example shown in Figure 5.13.

   (b) There exist overlaps with other photoelectron or Auger electron peaks. These can induce an asymmetry on either side of the main peak of interest since the overlapping peaks may occur at any $B.E._{XPS}$ value.

   (c) Insufficient charge neutralization is applied (insulators) or grounding is ensured (all sample types) during analysis. In mild cases, this will introduce an asymmetry onto the low $B.E._{XPS}$ side of all core-level peaks, with an example shown in Figure 4.4.

Case (a) can be dealt with by applying an asymmetry to the photoelectron peak of interest (available in most XPS software packages). This can be accomplished by using the Doniach–Sunjic line shape approach (Doniach and Sunjic, 1970) where an asymmetry is systematically applied to the high $B.E._{XPS}$ side of an initially Lorentzian peak until an effective fit is noted. Case (b) requires knowledge of the shape of the overlapping photoelectron peak. This can be derived from photoelectron peaks from other levels of the same element. Case (c) can and should be removed at the analysis stage since the distortion to the photoelectron peak may hide other important information (see Section 4.1.4). Note: The small variations in $B.E._{XPS}$ values that often remain can easily be corrected for at the processing stage.

Only once the asymmetry is removed or accounted for can any over-lapping peak be examined. This can be carried out using the following procedures:

(a) Derive a mechanism that is of physical significance to the system under study.

(b) Examine related spectra, that is, those collected under different analytical conditions such as takeoff angle for systematic patterns that would be expected from the system under study.

(c) Constrain the FWHM of the photoelectron peak of interest to those of surrounding related peaks. Note: All bulk photoelectron peaks from the level of the same element present within the bulk tend to exhibit similar FWHM. Elements present at the surface tend to exhibit larger FWHM than those present in the bulk (this results from the less stable surface environment, which in turn shortens core hole lifetimes).

(d) Minimize the number of contributions necessary to fit the spectra observed. This minimizes the possible number of outcomes, thereby enhancing the physical significance of the result obtained.

(e) Use of *principal component analysis* or *factor analysis* (Gilbert et al., 1982). This statistical procedure attempts to filter out all possible contributions based on the data set available. Although this can prove to be a somewhat complex task, effective results can be obtained by the trained user.

Note: Curve fitting can only be undertaken following the application of some background subtraction routine as discussed in Section 4.2.3. Examples of curve fitted O-1s spectra are shown in Figure 5.20. These spectra were fitted by

(a) Accessing the minimum possible number of environments in which oxygen can reside in YBCO.

(b) Evaluating the spectral variations with the takeoff angle and how these match the known structure.

(c) Fixing the FWHM and $B.E._{XPS}$ of the peaks assigned to oxygen in different environments.

(d) Comparing the assignments made with those reported in the available literature.

Further details on these results along with a full list of the derived $B.E._{XPS}$ values can be found in Section 6.1.4.

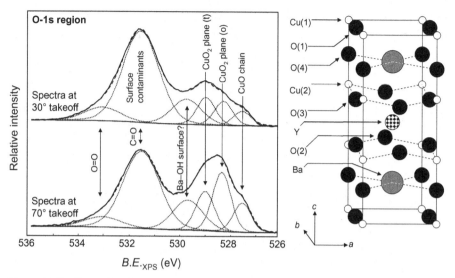

Figure 5.20. Curve fitted O-1s spectra from a tetragonal YBCO (YBCO[t]) terminated 20-nm-thick orthorhombic YBCO (YBCO[o]) film on MgO. On the right is the associated ball and stick diagram of the structure of YBCO(o) (the *a–b* plane represents the surface plane). The spectra, collected at room temperature as described in Section 6.1.4, are represented by the solid lines. These were fitted using symmetric peaks (90% Gaussian, 10% Lorentzian), represented by the dotted lines, after a Shirley background subtraction had been applied. The dashed lines represent their sum. Peak assignments were made on the basis of the variations noted with takeoff angle and those in various literature studies (for details, see Section 6.1.4). The CuO chain refers to oxygen in O(1) sites, CuO_2 plane(o) refers to O^{2-} in O(2) and O(3) sites in YBCO(o), and CuO_2 plane(t) refers to O^{2-} in O(2) and O(3) sites in YBCO(t). *Note*: Ba–OH is a tentitive assignment. Reproduced with permission from van der Heide (2006).

5.2 SUMMARY

Photoelectron spectra comprise intensity (I) versus binding energy ($B.E._{XPS}$) plots. The parameters that govern photoelectron intensities are discussed in Chapter 4. The parameters that govern photo $B.E._{XPS}$ values are discussed in Chapters 2 and 5. In Chapter 2, it was shown that $B.E._{XPS}$ values are a function of the number of protons within the nucleus (this defines the element) and the proximity of these electrons to the nucleus (this is defined by the stationary state) with a roughly $1/r^2$ dependence (r is the radius).

Subtle shifts in $B.E._{XPS}$ values ($\Delta B.E._{XPS}$) along with the introduction or otherwise of additional features in the XPS spectra are, however, noted as a free atom becomes ionized and/or becomes bound to other elements. Since these shifts/features can be correlated to the type of environment the atom is bound in, they are useful in revealing the speciation of the respective elements. Indeed, understanding their underlying cause is useful in situations where little is known about the system under study.

The underlying cause of the satellites often noted has been ascribed to those arising from

(a) *Initial State Effects*: These describe the effect induced by the bonding that occurs with other atoms/ions. Note: Although only valence electrons partake in bonding, all electrons (valence and core electrons) experience the change in electron density induced as described by the charge potential model. Many $\Delta B.E._{XPS}$ have thus been related to such parameters as

   (i) *The EN of Neighboring Atoms/Ions*: Note: This only applies for systems displaying very similar chemistries.

   (ii) *Bond Distance/Atomic Radii*: These provide a more direct measure of electron density, but accurate values can be hard to come by.

(b) *Final State Effects*: These describe the effect induced by the perturbation of the electronic structure resulting from photo-electron emission, particularly when core levels are involved. Since such effects also depend on the initial electronic structure (that from bonding), they too can be useful in revealing the original speciation of the photoelectron emitting atom/ion.

Initial state effects can be further subdivided as those arising from

(a) *Ground-state polarization* (bonding) or more precisely

   (i) Interatomic effects (those from neighboring atoms/ions)

   (ii) Intra-atomic effects (those from within the atom/ion)

(b) *Spin orbit splitting*

Final state effects can be further subdivided as those arising from

(a) *Excited state polarization* of the photoelectron emitting atom/ion. In other words, the core hole remaining will modify $B.E._{XPS}$ values derived from the departing photoelectrons as a result of

   (i) *Coulombic* (intra-atomic) effects

   (ii) *Mulitiplet splitting* (intra-atomic) effects

(b) *Rearrangement* effects, that is, core hole-induced excitation of valence electrons followed in some cases by relaxation into different states. Such process can introduce new features into the XPS spectra. Examples include

(i)   Shake-up satellites

(ii)   Shake-off satellites

(iii) Plasmon features

(iv) Auger peaks

(v)   Peak asymmetry

To complicate matters, final state effects also vary as a function of the core hole lifetime. Note: These decay through Auger processes and can be related to the irreducible photoelectron peak width (a result of the uncertainty principle). As a result, final state effects are always active. Their effects, however, may or may not be apparent.

CHAPTER 6

SOME CASE STUDIES

6.1 OVERVIEW

The purpose of this chapter is to illustrate how various technological questions of interest can be studied using non-synchrotron-based instruments (synchrotron-based instruments would have provided the same answers but to improve sensitivity and energy resolution). To retain the concise nature of this text, a very limited number of examples were chosen. The examples presented cover

(a) The assessment of X-ray photoelectron spectroscopy (XPS) for studying I_2 impregnation of carbon nanotubes as potential vessels for computer tomography (CT) imaging

(b) The examination and tabulation of the $B.E._{XPS}$ ($B.E._{XPS}$ is defined in Section 1.4) values of the group IIA–IV metals, their oxides and their carbonates.

(c) The study of various mixed metal oxides of interest as cathode materials in the development of solid oxide fuel cells (SOFCs).

(d) The investigation of the spectra noted from all the elements present within $YBa_2Cu_3O_{7-\delta}$ in its various phases as defined by δ.

X-ray Photoelectron Spectroscopy: An Introduction to Principles and Practices, First Edition. Paul van der Heide.

In all cases, a monochromatic Al-$K\alpha$ X-ray source was used to generate the signals of interest. Spectra were collected using concentric hemispherical analyzers (CHAs) operated in the constant analyzer energy (CAE)/fixed analyzer transmission (FAT) mode. To deal with insulating samples, low-energy electron charge neutralization was implemented along with, in some cases, a low-energy ion beam. Heating was carried out either within the analysis chamber or in an attached reaction chamber when necessary.

6.1.1 Iodine Impregnation of Single-Walled Carbon Nanotube (SWNT)

SWNTs, as the name suggests, are a tubular structure composed of a single atomic layer of carbon. Carbon nanotubes also come in the form of multiwalled carbon nanotubes (MWNTs). SWNTs can be conceptualized as a single-atom thick layer of graphite, otherwise called graphene, wrapped into a tubular shape of ~1-nm diameter. Their length can extend over many micrometers.

Carbon nanotubes, not to be confused with carbon fiber, are an allotrope of carbon that exhibits many novel properties. For example, these are the strongest and stiffest materials yet discovered to date in terms of tensile strength and elastic modulus. SWNTs also exhibit interesting electrical properties; that is, they have a bandgap that can extend from 0 to 2 eV. Although MWCTs do not share these same electrical properties, they are more impervious to chemical attack than SWNTs. Indeed, covalent functionalization of SWNTs tends to break the C–C bonds leaving holes in the tube walls.

Filling of SWNTs has attracted significant attention since the optical, magnetic, and/or electrical properties of the filler can be protected by the carbon outer layer. In addition, the carbon outer layer can act as a barrier to the potentially hazardous properties of the filler to the outside world. One area is in the medical field, in which toxic ions (e.g., Gd^{3+} for MRI) or molecules (e.g., I_2 for CT imaging) are of significant interest. More specifically, the question is whether these can be effectively encapsulated within SWNTs, and if so, can this be efficiently identified using some microanalytical technique?

One method of producing SWNT is to expose CO at a pressure (Pco) of 30–50 atm to an iron catalyst held at high temperature (>700°C) (Smalley, 2001). This results in a hemisphere of graphite that forms on the iron catalyst. When this lifts off, this becomes one end of the SWNT, which grows thereafter to lengths up to 5 μm. Fe then acts to close off the other end. The SWNTs prepared via this method are referred to as

HiPco SWNTs. These are preferred for medical applications due to their greater uniformity.

Loading of these SWCTs with I_2 can be carried out via sublimation (the belief is that the I_2 enters through the Fe capped end). SWNTs, however, tend to aggregate into large bundles, thereby introducing the possibility of trapping (intercalating) hydrophobic molecules such as I_2 within the spaces between individual SWNTs. One method to remove the intercalated I_2 is to debundle these tubes. This can be carried out through reduction with Na metal in dry tetrahydrofuran ($Na°/THF$) with the external I_2 washed away as I^-. These can then be dried into flakes in the millimeter range. Further details on the synthesis can be found elsewhere (Kissell, 2006).

High-resolution transmission electron microscopy (HRTEM) is one method that has been used for characterizing filled SWNTs. Although effective, particularly when scanning transmission electron microscopy (STEM) and/or electron energy loss spectroscopy (EELS) are applied, this technique only analyzes a small portion of a bulk sample (see Appendix G for a discussion on transmission electron microscopy [TEM]). This case study illustrates the potential of XPS in determining the amount and location of I_2 on and within large bundles of SWNTs.

Representative examples of XPS survey spectra collected from both separate bundles of the untreated and reduced I_2 filled SWNTs are shown in Figure 6.1. This is shown to illustrate all the peaks present; that is, only those from carbon and iodine are noted with a minor contribution from oxygen also present (that of the support is completely quenched). Representative high-resolution spectra collected over the

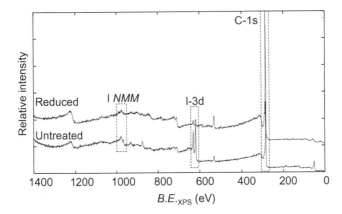

Figure 6.1. XPS spectra collected over the 0- to 1400-eV $B.E._{XPS}$ range from separate bundles of untreated and reduced I_2 filled SWNTs. The regions of interest are signified. The only other signal of note is the weak O-1s photoelectron peak at ~530 eV.

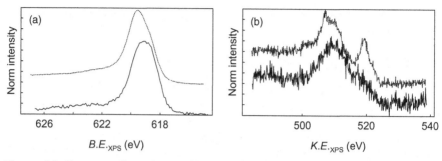

Figure 6.2. Spectra collected over (a) the I-3d$_{5/2}$ photoelectron region and (b) the I-$M_4N_{45}N_{45}$ Auger electron region. These were collected from separate bundles of untreated (dashed lines) and reduced I$_2$ filled SWNTs (solid lines). Peak intensities were normalized and then the spectra were shifted along the intensity axis for the sake of clarity. Reproduced with permission from Kissell et al. (2006).

I-3d$_{5/2}$ photoelectron and the I-$M_4N_{45}N_{45}$ Auger electron regions, respectively, are shown in Figure 6.2a,b. That collected over the C-1s region is shown in Figure 5.16. To aid in the assignment of the recorded photoelectron and X-ray-induced Auger electron emissions, as-received and reduced bundles of SWNTs were examined. SWNT bundles (in the form of dried flakes) were supported by pressing into indium foil as described in Section 4.1. Indeed, several methods were examined with this providing the best results.

Quantification of the XPS spectra revealed I$_2$ was present at ~5.3 atomic % on/within the untreated SWNTs. This agreed well with the mass gain observed during the filling process. Upon reduction, the amount of I$_2$ decreased to ~2.8 atomic %. This value, which approximates that an average of ~3 iodine atoms exist per nanometer of SWNT, is consistent with inductively coupled plasma atomic emission spectrocsopy (ICP-AES) studies carried out on the THF/water filtrate (this was carried out on the remaining NaI after the THF was thermally removed). Note: Sodium was not observed in the XPS spectra.

The relatively broad I-3d$_{5/2}$ photoelectron peak noted at a $B.E._{XPS}$ of 619.5 eV from the untreated SWNTs is consistent with that of I$_2$, as opposed to polyiodide chains (I$_3$ or I$_5$). This decreased slightly on reduction to a $B.E._{XPS}$ of 619.1 eV. In addition, no other discrete peaks were noted over the I-3d$_{5/2}$ or C-1s regions (only the C-1s 284.6-eV peak was noted along with the π to π* shake-off structure expected from graphite-like structures). This suggests that the I$_2$ introduced did not react with the SWNTs; that is, C–I bonds do not appear to have been formed. No definitive information on the location of the I$_2$ (internal or external to the SWNTs) could, however, be conclusively inferred from any of the core-level photoelectron emissions recorded.

As noted in Figure 6.2b, significant spectral changes were evident over the I-$M_4N_{45}N_{45}$ Auger region. This is primarily evident in the disappearance of the two prominent peaks noted at a $K.E._{XPS}$ ($K.E._{XPS}$ is defined in Section 1.4) of 507.5 and 519 eV for the untreated SWNTs on reduction. Although the peak at 519 eV is consistent with previous studies carried out on I_2 (same values are listed in Briggs and Seah, 1990), that at 507.5 eV overlaps with the O-KLL region. As a result, the 507.5-eV peak could not be conclusively assigned even though following the trends exhibited by all of the other iodine signals, as well as those expected. Note: As can be seen in Figure 6.1, the O-1s peak at ~530 eV increases in intensity on reduction, that is, it exhibits the opposite trend to that exhibited by the 507.5-eV peak, as well as all other iodine signals.

The I-$M_4N_{45}N_{45}$ Auger spectra from the reduced SWNT exhibited broader structures situated at $K.E._{XPS}$ values of ~510 and ~517 eV. These are believed to be representative of I_2 situated within the SWNTs (inferred on the basis that reduction in Na°/THF removes any intercalated I_2). The slight decrease in the $K.E._{XPS}$ of this peak with respect to literature values for I_2 (517 eV vs. 519 eV from Briggs and Seah, 1990), as well as the slight decrease in the $B.E._{XPS}$ of the I $3d_{5/2}$ peak (619.5–619.1 eV as can be seen in Fig. 6.2a), is believed to be due to the different environment experienced by I_2 when contained with these SWNTs.

6.1.2 Analysis of Group IIA–IV Metal Oxides

The aim of this study was to reexamine the $B.E._{XPS}$ shifts exhibited by O and the group IIA, IIIA, and IVA elements on oxide and carbonate formation, such that a unified and self-consistent $B.E._{XPS}$ data set could be derived. These elements/substrates were chosen since

(a) No obvious trend is noted on comparing the O-1s $B.E._{XPS}$ values for the group IA–VA oxides from that available in the literature, a fact compromised by the spread in reported values, even for the same oxide; that is, values from 528.2 to 531.5 eV exist for SrO.

(b) The $B.E._{XPS}$ values of the heavier group IA and IIA elements decrease on oxidation, while except for La, $B.E._{XPS}$, values of the group III, IV, and VA elements increase. In addition, highly simplistic but conflicting arguments based on superionicity and partial covalency, along with work function, have been used to explain these trends. Further discussion on these can be found in Section 5.1.1.2.2.1.

(c) With the exception of La, photoelectrons from these elements do not suffer strong relaxation or multiplet splitting (La core levels exhibit satellites), while polarization is assumed minimal. The latter is inferred on the basis of

(i) The similar $B.E._{XPS}$ shifts ($\Delta B.E._{XPS}$) exhibited by the different Ba core levels

(ii) The consistency of the trends within these groups expected on the basis of initial state effects

With the exception of Sr, Ba, and La, all samples consisted of 99.99% or higher purity foils. These were cleaned with acetone and methanol prior to analysis. Sr, Ba, and La were cut from 99.9% purity ingots, which were prepared and transferred in an Ar environment to avoid the formation of carbonates (this occurs on exposure to atmospheric gases). Clean elemental surfaces were then produced by sputtering with 1-keV Ar^+ followed by annealing at 600°C in ultrahigh vacuum (UHV). Sputter–anneal cycles were carried out until the C-1s impurity peak at ~285 eV fell below acceptable limits. All samples were acquired from Johnson Matthey PLC.

Oxides of Ca, Sr, Ba, and La were then formed through heating the respective carbonates *in situ* or within an attached reaction chamber (this induces the controlled decomposition of the carbonates to CO_2 gas and the oxide of interest). Though this is accelerated when carried out in O_2, UHV was preferred since this allowed for all spectral variations to be followed. Carbonates of these elements were formed through exposure of the cleaned surfaces to CO_2 or the atmosphere. Oxides of all other elements were produced through exposure of the cleaned surfaces to O_2. Exposure to the various gases was carried out in an attached reaction chamber.

Note: La was the only group IIIA element to exhibit carbonate formation. Carbonates are revealed through the growth of a 291.4-eV peak in the C-1s region. This was also the only element aside from Ca, Sr, and Ba to exhibit the formation of shake-up satellites (these are situated at ~4.6 eV above the La-3d peaks) and the only element to exhibit decreasing $B.E._{XPS}$ values on oxide formation. Carbonate formation was not noted on any of the group IVA elements.

Representative examples of spectra acquired from Sr and Ba are shown in Figures 6.3a–d and 6.4a–d. Those from La and Ti are shown in Figure 6.5a–d. The latter are shown since they represent the extremes within the group IIIA and IVA elements in their rate of oxidation (Ti fastest and La slowest), and the $\Delta B.E._{XPS}$ values (Ti exhibited the largest increase while La displayed a decreasing $B.E._{XPS}$ on oxidation).

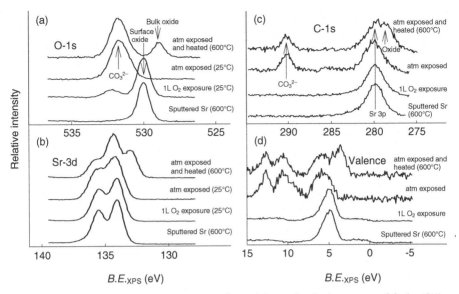

Figure 6.3. Raw photoelectron spectra collected from the Sr ingot over (a) the O-1s region, (b) the Sr-3d region, (c) the C-1s region, and (d) the valence region. These were collected under the conditions listed with heating applied *in situ*. The spectra are rescaled and offset along the intensity axis to more clearly illustrate the variations. Reproduced with permission from van der Heide (2006).

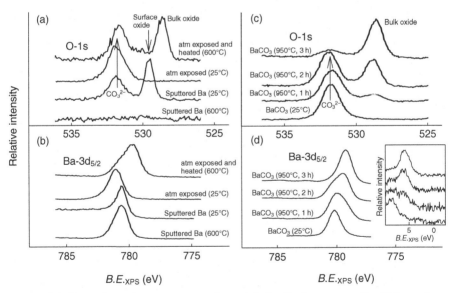

Figure 6.4. Raw photoelectron spectra from (a) the Ba ingot over the O-1s region, (b) the Ba ingot over the Ba-3d$_{5/2}$ region, (c) the BaCO$_3$ powder over the O-1s region, and (d) the Ba powder over the Ba-3d$_{5/2}$ region. These were collected under the conditions listed. The spectra are rescaled and offset along the intensity axis to more clearly illustrate the variations. In the inset of panel (d) are shown valence band spectra from BaCO$_3$ powder collected under the same conditions and in the same order as in panel (d). Reproduced with permission from van der Heide, 2006.

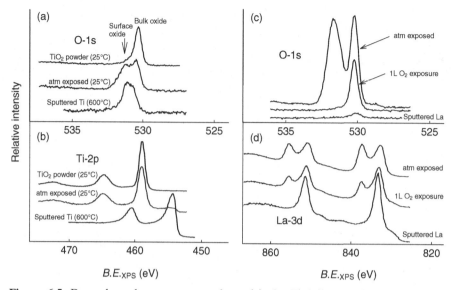

Figure 6.5. Raw photoelectron spectra from (a) the Ti foil over the O-1s region, (b) the Ti foil over the Ti-2p region, (c) the La ingot over the O-1s region, and (d) the La ingot over the La-3d$_{5/2}$ region. These were collected under the conditions listed with spectra rescaled and offset along the intensity axis to more clearly illustrate the variations. Reproduced with permission from van der Heide (2006).

The $B.E._{XPS}$ values were referenced, when necessary, using one or more applicable methods. In the case of metallic substrates, the Fermi edge was used, that is, affixed to a $B.E._{XPS}$ value of 0.0 eV. In the formation of surface oxides/carbonates, an unaffected peak was used as the internal reference (typically a bulk metal peak). In the case of bulk oxides (those where the Fermi edge is contained within the bandgap), the bulk O-1s peak was used as an internal reference. The position of this O-1s peak was derived by heating the respective oxide *in situ* until fully conductive, that is, up to 950°C. In the case of the carbonates, the bulk C-1s peak indicative of the carbonate (that located at 290.6 eV on $CaCO_3$ at 950°C) was used. Other referencing procedures such as the use of the C-1s adventitious peak (that adsorbed at room temperature) were not used since a range of values from 284.4 to 285.2 eV exist, and carbon speciation was found to be influenced by the preparation procedures used. The $B.E._{XPS}$ values derived from the raw spectra are listed in Table 6.1. Those noted after curve fitting are listed in Table 6.2.

The O-1s $B.E._{XPS}$ values collected from a particular substrate were then plotted against the EN of the most EN neighboring atom. This is shown in Figure 5.6b for O on La. Attempts to extend this relation to all the surfaces analyzed proved less effective even when separating the O-1s signals into their groups, that is, bulk oxides and surface oxides.

TABLE 6.1   C-1s and O-1s $B.E._{XPS}$ Values Recorded from Various Group IIA–IVA Elemental Surfaces along with the Most Prominent Metal Core Level Are Tabulated Below

Surface	Sample Type	C-1s	O-1s	Metal Core Level (Level)	$B.E._{XPS}$ Reference (Source)
Ca	Metal	<DL	530.2	346.5 ($2p_{3/2}$)	E_F (PvdH)
	Metal		530.1	345.9 ($2p_{3/2}$)	E_F (Sol)
	Metal		NL	345.9–346.8 ($2p_{3/2}$)	V (Phi)
	Film		NL	345.9 ($2p_{3/2}$)	V (NIST)
CaO	On metal	<DL	529.9(surface)	346.8 ($2p_{3/2}$)	O (PvdH)
	On metal	<DL	528.8	346.1 ($2p_{3/2}$)	O (PvdH)
	Powder	285.2	529.0	346.5 ($2p_{3/2}$)	Δ,O(PvdH)
	On metal	285.0*	528.9	346.0 ($2p_{3/2}$)	C-1s* (Sol)
	Various		529.4–531.3	346.2–346.8 ($2p_{3/2}$)	V (Phi)
	Various		529.4–531.3	346.1–347.3 ($2p_{3/2}$)	V (NIST)
CaCO₃	On metal	290.6	532.7	347.6 ($2p_{3/2}$)	O (PvdH)
	Powder	290.6*	532.5	347.5 ($2p_{3/2}$)	O,C-1s* (PvdH)
	Powder	289.7, 285.0*	531.6	347.3 ($2p_{3/2}$)	C-1s* (Sol)
	Various		531.4	346.4–347.3 ($2p_{3/2}$)	V (Phi)
	Various		NL	346.7–347.0 ($2p_{3/2}$)	V (NIST)
Sr	Metal	<DL	530.0	133.1 ($3d_{5/2}$)	E_F (PvdH)
	Metal		530.0	133.7 ($3d_{5/2}$)	E_F (Sol)
	Metal		NL	134.3–134.7 ($3d_{5/2}$)	V (Phi)
	Metal		NL	134.2–134.4 ($3d_{5/2}$)	V (NIST)
SrO	On metal	<DL	530.3(surface)	133.4($3d_{5/2}$)	O (PvdH)
	On metal	<DL	528.5	132.1 ($3d_{5/2}$)	O (PvdH)
	Powder	285.0	528.6	132.3 ($3d_{5/2}$)	Δ,O (PvdH)
	On metal	285.0*	528.2	132.4 ($3d_{5/2}$)	C-1s* (Sol)
	Various		530.5	135.1–135.7 ($3d_{5/2}$)	V (Phi)
	Various		530.2–530.5	132.8–135.3 ($3d_{5/2}$)	V (NIST)
SrCO₃	On metal	290.0*	532.4	133.8 ($3d_{5/2}$)	O (PvdH)
	Powder	290.0*	532.4	133.8 ($3d_{5/2}$)	O,C-1s* (PvdH)
	Powder	289.5, 285.0*	531.5	133.4 ($3d_{5/2}$)	C-1s* (Sol)
	Various		NL	133.1–133.6 ($3d_{5/2}$)	V (Phi)
	Powder		NL	132.9 ($3d_{5/2}$)	V (NIST)
Ba	Metal	<DL	529.5	780.5 ($3d_{5/2}$)	E_F (PvdH)
	Metal		529.8	780.8 ($3d_{5/2}$)	E_F (Sol)
	Metal		NL	780.1–780.6 ($3d_{5/2}$)	V (Phi)
	Various		NL	779.3–780.7 ($3d_{5/2}$)	V (NIST)
BaO	On metal	<DL	529.4(surface)	780.7 ($3d_{5/2}$)	O (PvdH)
	On metal	<DL	528.0	779.1 ($3d_{5/2}$)	O (PvdH)
	Powder	285.5	528.0	779.3 ($3d_{5/2}$)	Δ,O (PvdH)

(Continued)

TABLE 6.1 (*Continued*)

Surface	Sample Type	C-1s	O-1s	Metal Core Level (Level)	$B.E._{XPS}$ Reference (Source)
	On metal	285.0*	527.6	778.8 ($3d_{5/2}$)	C-1s* (Sol)
	Various		528.3	779.1–779.9 ($3d_{5/2}$)	V (Phi)
	Various		528.3–530.2	779.1–779.9 ($3d_{5/2}$)	V (NIST)
$BaCO_3$	On metal	290.0	532.0	780.8 ($3d_{5/2}$)	O (PvdH)
	Powder	290.0*	531.8	780.7 ($3d_{5/2}$)	O,C-1s* (PvdH)
	Powder	289.1, 285.0*	531.2	779.6 ($3d_{5/2}$)	C-1s* (Sol)
	Various		531.4	779.7–781.2 ($3d_{5/2}$)	V (Phi)
	Various		531.1–532.2	779.8–779.9 ($3d_{5/2}$)	V (NIST)
Sc	Metal	281.7	531.5	398.6 ($2p_{3/2}$)	E_F (PvdH)
	Metal		NL	398.6–399.0 ($2p_{3/2}$)	V (Phi)
	Metal		NL	398.3–398.8 ($2p_{3/2}$)	V (NIST)
Sc_2O_3	On metal	284.9	529.9, 531.3	401.6 ($2p_{3/2}$)	$O^{531.5}$ (PvdH)
	Various		530.0	401.5–402.5 ($2p_{3/2}$)	V(Phi)
	Various		529.4–530.1	401.3–403.4 ($2p_{3/2}$)	V(NIST)
Y	Metal	281.2	531.0	156.0 ($3d_{5/2}$)	E_F (PvdH)
	Metal		NL	155.5–156.0 ($3d_{5/2}$)	V (Phi)
	Metal		NL	155.8–156.1 ($3d_{5/2}$)	V (NIST)
Y_2O_3	On metal	286.3	529.5, 531.0	157.5 ($3d_{5/2}$)	$O^{531.0}$ (PvdH)
	Various		NL	156.4–157.0 ($3d_{5/2}$)	V (Phi)
	Various		529.3–530.2	156.6–158.6 ($3d_{5/2}$)	V (NIST)
La	Metal	282.6	530.9	835.6 ($3d_{5/2}$)	E_F (PvdH)
	Metal		NL	835.6–836.2 ($3d_{5/2}$)	V (Phi)
	Metal		NL	835.9 ($3d_{5/2}$)	V (NIST)
La_2O_3	On metal	<DL	529.0, 530.6	835.3 ($3d_{5/2}$)	$O^{530.6}$ (PvdH)
	Powder	<DL	529.2, 530.6, 533.5	835.0 ($3d_{5/2}$)	$\Delta,O^{530.6}$ (PvdH)
	Various		528.6	833.5–835.2 ($3d_{5/2}$)	V (Phi)
	Various		528.4–532.8	832.5–835.1 ($3d_{5/2}$)	V (NIST)
$La_2(CO_3)_3$	On metal	291.4, 286.1	530.6, 532.7	833.5 ($3d_{5/2}$)	$\Delta,O^{530.6}$ (PvdH)
Ti	Metal	282.2	531.8	454.0 ($2p_{3/2}$)	E_F (PvdH)
	Metal		NL	453.7–454.2 ($2p_{3/2}$)	V (Phi)
	Metal		NL	453.8–454.3 ($2p_{3/2}$)	V (NIST)
TiO_2	On metal	<DL	530.2, 531.8	459.1 ($2p_{3/2}$)	$O^{531.8}$ (PvdH)
	Powder	285.2	530.2, 531.8	459.0 ($2p_{3/2}$)	$\Delta,O^{531.8}$ (PvdH)
	Various		529.9	458.7–459.3 ($2p_{3/2}$)	V (Phi)
	Various		529.7–530.2	458.5–458.9 ($2p_{3/2}$)	V (NIST)
Ti_2O_3	On metal	<DL	531.0, 531.8	457.9 ($2p_{3/2}$)	$O^{531.8}$ (PvdH)
	Oxidized NiTi		529.6	456.9 ($2p_{3/2}$)	V (NIST)
Zr	Metal	<DL	531.0	179.1 ($3d_{5/2}$)	E_F (PvdH)
	Metal		NL	178.7–178.9 ($3d_{5/2}$)	V (Phi)
	Metal		NL	178.3–179.3 ($3d_{5/2}$)	V (NIST)

TABLE 6.1 (*Continued*))

Surface	Sample Type	C-1s	O-1s	Metal Core Level (Level)	$B.E._{\mathrm{XPS}}$ Reference (Source)
ZrO_2	On metal	<DL	530.2, 531.0	182.4 ($3d_{5/2}$)	$O^{531.0}$ (PvdH)
	Various		530.2–530.9	182.1–182.6 ($3d_{5/2}$)	V (Phi)
	Various		529.9–531.3	182.2–184.0 ($3d_{5/2}$)	V (NIST)

All values are in electronvolt with respect to E_F. Various reference procedures were used as indicated in the key. These data were collected from the various clean metal surfaces and following various exposures as outlined in, and with permission from, van der Heide (2006).

Keys to Table 6.1

PvdH	Data from van der Heide (2006).
Sol	Data from Sosulnikov and Teterin (1992).
Phi	Data from Moulder et al. (1992).
NIST	Data from Wagner et al. (2003).
<DL	Peak is below the instrument detection limit.
NL	These values are not listed in the respective databases.
E_F	Energy scale is defined by setting the Fermi edge to 0.0 eV (conductive matrices only).
Δ	Energy scale is defined by heating the sample until conductive (400–950°C).
O	Energy scale is defined by setting O-1s $B.E._{\mathrm{XPS}}$ to the $B.E._{\mathrm{XPS}}$ derived in O adsorption experiments carried out on conductive metal surfaces (superscript refers to the peak $B.E.$ used).
*	Energy scale is defined by setting the C-1s adventitious peak to 285 eV or by setting the C-1s carbonate peak to the listed $B.E._{\mathrm{XPS}}$ (290.0–290.6 eV depending on the sample).
V	Various $B.E._{\mathrm{XPS}}$ referencing procedures used as outlined within Moulder et al. (1992) and Wagner et al. (2003).

The O-1s $B.E._{\mathrm{XPS}}$ values from all of the bulk oxides do, however, appear to scale with $1/d$, where d is the interatomic distance. This is illustrated in Figure 5.7. The same is noted for the $\Delta B.E._{\mathrm{XPS}}$ values observed from the metallic counterparts on oxidation. This and the decreasing $B.E._{\mathrm{XPS}}$ values exhibited by Ca, Sr, and Ba on oxidation can be explained within the charge potential model as resulting from competing initial state inter- and intra-atomic effects and appears consistent with partial covalency arguments utilizing Madulung potentials (see Section 5.1.1.2.2.1).

6.1.3 Analysis of Mixed Metal Oxides of Interest as SOFC Cathodes

Mixed metal oxides are of technological interest since their properties lend themselves well to the development of electrocatalysts, memory cells, optical wave guides, as well as substrates for epitaxial superconductor growth. This arises from the high stability of these oxides under various reactive environments (a result of their structure), and the ability to tailor the properties of these oxides to those required.

TABLE 6.2   O-1s $B.E._{XPS}$ Values Recorded from Various Group IIA–IVA Elemental Surfaces along with the Most Prominent Metal Core Level during Oxide or Carbonate Formation are Tabulated Below

Core Level	O-1s	Surface	Assignment Ordering
O-1s	528.8/530.2/531.5/~533	Ca	Bulk oxide/surface oxide/ carbonate/Ca–O=O
O-1s	528.5/530.0/531.4/~533	Sr	Bulk oxide/surface oxide/ carbonate/Sr–O=O
O-1s	528.0/529.5/531.0/~533	Ba	Bulk oxide/surface oxide/ carbonate/Ba–O=O
O-1s	529.9/531.5	Sc	Bulk oxide/surface oxide
O-1s	529.5/531.0	Y	Bulk oxide/surface oxide
O-1s	529.0/530.9/532.0/~533	La	Bulk oxide/surface oxide/ carbonate/La–O=O
O-1s	530.2/531.8	Ti	Bulk oxide/surface oxide
O-1s	530.2/531.0	Zr	Bulk oxide/surface oxide
Ca-2p$_{3/2}$	346.5/346.1/347.5	Ca	Metal/bulk oxide/carbonate
Sr-3d$_{5/2}$	133.1/132.1/133.8	Sr	Metal/bulk oxide/carbonate
Ba-3d$_{5/2}$	780.5/779.1/780.7	Ba	Metal/bulk oxide/carbonate
Sc-2p$_{3/2}$	398.6/401.6	Sc	Metal/bulk oxide
Y-3d$_{5/2}$	156.0/157.5	Y	Metal/bulk oxide
La-3d$_{5/2}$	835.6/835.3	La	Metal/bulk oxide/carbonate
Ti-2p$_{3/2}$	454.0/459.1	Ti	Metal/bulk oxide
Zr-2p$_{3/2}$	179.1/182.4	Zr	Metal/bulk oxide

Various reference procedures were used as indicated in the key. Due to asymmetry on the carbonate peak, the spectra were curve fitted using two contributions, that is, that from carbonates and the other from a weakly bound O=O component. The data were reproduced with permission from van der Heide (2006).

Electrocatalysts and ion transport membranes for oxygen separation (applications include sensors, O_2 pumps, membrane reactors, and SOFCs) rely on the adsorption and diffusion of oxygen (not to be confused with H-based SOFCs).

The rate at which oxygen is adsorbed from the ambient environment depends on, among other things, the structure and composition of the outermost surface of the oxide. As first discussed in Section 1.1, the structure and composition of a surface can differ from that of the bulk, a fact resulting from the abrupt termination of the lattice structure. Indeed, the decreased stability induced is often sufficient to bring about the segregation and/or structural distortion over this region. If apparent this will likely affect adsorbtion/desorption kinetics, all of which will be enhanced under the elevated temperatures at which such membranes typically operate.

The aim of this study is to use XPS to examine a selection of the $La_{1-x}Sr_xBO_{3-\delta}$-type oxides, where $La_{1-x}Sr_x$ represents the A-type cations in ABO_3 perovskite-type oxides, the B cations are various first-row transition elements and/or Ga, and δ represents the oxygen vacancies arising from the substitution of La by Sr (this has a lower oxidation state, hence the introduction of δ). Such substitution is desired since the introduction of oxygen vacancies enhance what is referred to as mixed electronic and ionic conductivity (MEIC). Substitution of Sr for La can affect the structural stability of these oxides, particularly the composition and speciation of the outermost surface.

Of particular interest are the spectral variations experienced by freshly prepared oxides on annealing in vacuum and under partial O_2 pressure environments at temperatures comparable to those akin to those used during operation. Spectra were thus collected from the as-received samples at room temperature, elevated temperatures, following *in situ* sputter and anneal cycles, and following exposure to various gases.

Representative O-1s, Sr-3d spectra collected from all as-received oxides at 45° takeoff angle are shown in Figure 6.6. Representative XPS spectra over the valence, O-1s, Sr-3d, and La-3d regions collected as a function of takeoff angle and temperature are shown in Figure 6.7a,b.

Two main features are observed in the O-1s spectra from $La_{0.5}Sr_{0.5}CoO_{3-\delta}$: one at 528.5 eV and one around 531–532 eV. The term

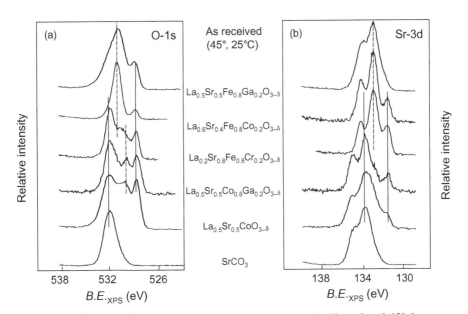

Figure 6.6. O-1s and Sr-3d photoelectron spectra at a takeoff angle of 45° from as-received polished surfaces of $La_{0.5}Sr_{0.5}CoO_{3-\delta}$, $La_{0.5}Sr_{0.5}Co_{0.8}Ga_{0.2}O_{3-\delta}$, $La_{0.5}Sr_{0.5}Fe_{0.8}Cr_{0.2}O_{3-\delta}$, $La_{0.4}Sr_{0.6}Fe_{0.8}Co_{0.2}O_{3-\delta}$, and $La_{0.5}Sr_{0.5}Fe_{0.8}Ga_{0.2}O_{3-\delta}$ pellets and of $SrCO_3$ powder. Figure reproduced with permission from van der Heide (2002).

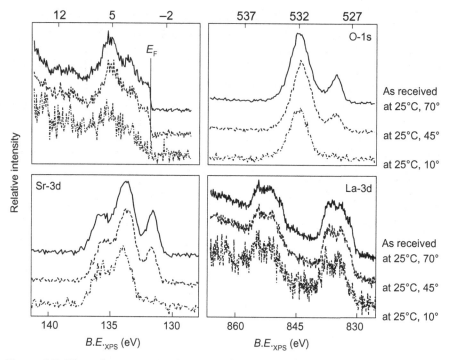

Figure 6.7. Photoelectron spectra of an as-received $La_{0.5}Sr_{0.5}CoO_{3-\delta}$ thin-film sample over the valence, O-1s, Sr-3d, and La-3d regions as a function of the takeoff angle over the 10–70° range at a temperature of 25°C and temperature and takeoff angle adjusted (these ranged between 25 and 600°C and 10 and 45°, respectively). The plots in the top left-hand corner of each are the respective valence band spectra. Figures reproduced with permission from van der Heide (2002).

features is used as opposed to peaks in cases where more than one contribution is believed to be responsible for the spectra (the terms *peaks* are only used when actual assignments to individual contributions can be made). The 531- to 532-eV feature decreases in intensity relative to that at 528.5 eV with increasing takeoff angle and increasing temperature and exhibits slight binding energy (*B.E.*) shifts. This signifies that surface states are responsible for the 531- to 532-eV feature, while bulk-bound states are responsible for the 528.5-eV feature.

Quantification of the $La_{0.5}Sr_{0.5}CoO_{3-\delta}$ thin-film sample revealed a surface enriched in Sr by a factor of ~1.2–2.0 times. Representative values are listed in Table 6.3. A Ga enrichment by a factor of ~1.5 times, along with higher C concentrations were also noted.

Curve fitting was then carried out on the assumption that all spectra could be systematically reproduced using a select (minimum) number

TABLE 6.3   XPS-Derived Composition of the $La_{0.5}Sr_{0.5}CoO_{3-\delta}$ Thin-Film Sample as a Function of the Takeoff Angle; This Sample Was Previously UHV Annealed at 800°C and Then Exposed to the Atmosphere

	C	O	Co	Sr	La	AB Stoichiometry
10°	78.2	16.3	1.3	3.2	1.0	$La_{0.4}Sr_{1.1}Co_{0.5}$
15°	78.5	16.4	1.1	3.1	0.9	$La_{0.4}Sr_{1.2}Co_{0.4}$
25°	74.4	19.1	1.5	3.7	1.3	$La_{0.4}Sr_{1.1}Co_{0.5}$
45°	64.1	26.5	2.6	5.0	1.8	$La_{0.4}Sr_{1.0}Co_{0.6}$
70°	52.3	34.4	4.6	6.0	2.8	$La_{0.4}Sr_{0.9}Co_{0.7}$

The table is reproduced with permission from van der Heide (2002).

of contributions whose *B.E.* and full width at half maxima (FWHM) remain fixed for all oxides.

In the case of the O-1s spectra, curve fitting was carried out as follows. A contribution was introduced at 528.5 eV since a symmetric feature is noted in the original spectra at this position on various clean oxides. The fact that this feature increases in intensity with increasing takeoff angle implies that this is due to photoemissions from O_2^- atoms situated within the bulk of these oxides. A second contribution was added at 529.8 eV since a symmetric feature was noted at this position on various cleaned surfaces. This is assigned to a surface O species since this grew with decreasing takeoff angle. Third and fourth contributions were added at 532.0 and 533.2 eV since this region could not be fitted with a single contribution of constant FWHM. These were assigned to surface contaminants since these were enhanced at lower takeoff angles and were removed, along with C, at elevated temperatures. Finally, a fifth contribution at 531.0 eV was added since a feature at this *B.E.* was also noted. This decreased in intensity with increasing takeoff angles indicative of surface-bound O.

FWHM values for the 528.5-, 529.8-, and 532-eV contributions were derived using their low energy slopes. The 528.5-eV contribution exhibited an FWHM of 1.07–1.14 eV, while those at 529.8 and 532 eV exhibited an FWHM of 1.35–1.5 eV. FWHM values used in curve fitting were henceforth fixed at 1.1 eV for the bulk oxide contribution and 1.4 eV for all others (these exhibited trends consistent with surface-bound O).

Curve fitting of the Sr-3d region was carried out using the same assumptions. FWHM varied between 1.1 eV for bulk-bound Sr and 1.2 eV for surface-bound Sr. These were defined from spectra in which one doublet predominated.

An example of the curve fitting carried out over the O-1s and Sr-3d regions is shown in Figure 6.8. All values derived are listed in

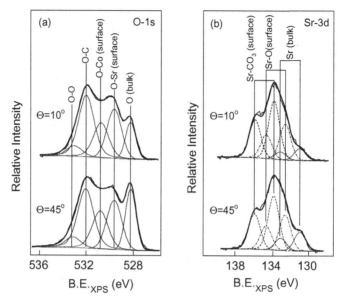

Figure 6.8. Curve fitted O-1s and Sr-3d spectra collected from the as-received $La_{0.5}Sr_{0.5}CoO_{3-\delta}$ at takeoff angles of 45° (those from Fig. 6.6) and 70°. The curve fitting procedure is described within Section 5.1.1.5. Figure reproduced with permission from van der Heide (2002).

TABLE 6.4   O-1s $B.E._{XPS}$ Values and FWHM Derived from Curve Fitting

Matrix	O-1s(I)	O(II)	O(III)	O(IV)	O(V)
$La_{0.5}Sr_{0.5}CoO_{3-\delta}$ (pellet)	528.5	—	531.3	—	533.5
$La_{0.5}Sr_{0.5}CoO_{3-\delta}$ (film)	528.5	529.8	531.0	532.0	533.2
$La_{0.5}Sr_{0.5}CoO_{3-\delta}$ (pellet)	528.5	529.9	531.3	532.1	533.1
$La_{0.6}Sr_{0.4}Co_{0.2}Fe_{0.8}O_{3-\delta}$ (pellet)	528.5	529.6	531.2	532.1	533.2
$La_{0.5}Sr_{0.5}Co_{0.8}Ga_{0.2}O_{3-\delta}$ (film)	528.5	529.8	531.2	532.2	533.1
$La_{0.5}Sr_{0.5}Co_{0.8}Ga_{0.2}O_{3-\delta}$ (pellet)	528.4	529.8	531.0	532.0	533.1
$La_{0.5}Sr_{0.5}Fe_{0.8}Ga_{0.2}O_{3-\delta}$ (film)	528.7	530.0	530.9	532.0	533.4
$La_{0.5}Sr_{0.5}Fe_{0.2}Ga_{0.2}O_{3-\delta}$ (pellet)	528.6	530.1	530.8	532.3	533.1
$La_{0.5}Sr_{0.5}Fe_{0.8}Cr_{0.2}O_{3-\delta}$ (pellet)	528.5	529.7	530.8	532.2	533.3

Peaks are arbitrarily grouped according to their $B.E._{XPS}$ values.

Tables 6.4–6.6. Assignments are listed in Table 6.7. Tables 6.4–6.7 were reproduced with permission from van der Heide (2002).

6.1.4   Analysis of YBCO and Related Oxides/Carbonates

One of the outstanding problems in condensed matter physics concerns the theory of superconductivity above 30 K (the Bardeen, Cooper, and

TABLE 6.5 Sr-3d$_{5/2}$ and C-1s $B.E._{\text{xps}}$ Values in Electronvolt Derived from Curve Fitting

Matrix	Sr-3d$_{5/2}$(i)	Sr-3d$_{5/2}$(ii)	Sr-3d$_{5/2}$(iii)	C-1s
La$_{0.5}$Sr$_{0.5}$CoO$_{3-\delta}$ (film)	131.6	132.9	134.0	284.9
La$_{0.5}$Sr$_{0.5}$CoO$_{3-\delta}$ (pellet)	131.7	132.8	133.8	284.8
La$_{0.6}$Sr$_{0.4}$Co$_{0.2}$Fe$_{0.8}$O$_{3-\delta}$ (pellet)	131.6	133.0	134.0	284.8
La$_{0.5}$Sr$_{0.5}$Co$_{0.8}$Ga$_{0.2}$O$_{3-\delta}$ (film)	131.5	132.9	133.8	284.8
La$_{0.5}$Sr$_{0.5}$Co$_{0.8}$Ga$_{0.2}$O$_{3-\delta}$ (pellet)	131.6	133.2	134.2	284.7
La$_{0.5}$Sr$_{0.5}$Fe$_{0.8}$Ga$_{0.2}$O$_{3-\delta}$ (film)	131.8	132.9	134.0	284.9
La$_{0.5}$Sr$_{0.5}$Fe$_{0.8}$Ga$_{0.2}$O$_{3-\delta}$ (pellet)	131.7	133.0	134.0	284.8
La$_{0.5}$Sr$_{0.5}$Fe$_{0.8}$Cr$_{0.2}$O$_{3-\delta}$ (pellet)	131.6	133.0	134.1	284.8

The Sr-3d$_{5/2}$ peaks are arbitrarily grouped as Sr(i), (ii), and (iii) groups.

TABLE 6.6 La-3d$_{5/2}$ $B.E._{\text{xps}}$ Values and Satellite Displacements Relative to the La-3d$_{5/2}$ Peaks as Derived from Curve Fitting

Matrix	La 3d$_{5/2}$(i)	La 3d$_{5/2}$(ii)	Satellites	
La$_{0.5}$Sr$_{0.5}$CoO$_{3-\delta}$ (film)	835.2	833.5	+2.1	+4.0
La$_{0.5}$Sr$_{0.5}$CoO$_{3-\delta}$ (pellet)	835.1	833.5	+2.1	+4.0
La$_{0.6}$Sr$_{0.4}$Co$_{0.2}$Fe$_{0.8}$O$_{3-\delta}$ (pellet)	834.7	833.0	+2.1	+4.0
La$_{0.5}$Sr$_{0.5}$Co$_{0.8}$Ga$_{0.2}$O$_{3-\delta}$ (film)	835.0	833.5		
La$_{0.5}$Sr$_{0.5}$Co$_{0.8}$Ga$_{0.2}$O$_{3-\delta}$ (pellet)	834.5	833.2		
La$_{0.5}$Sr$_{0.5}$Fe$_{0.8}$Ga$_{0.2}$O$_{3-\delta}$ (film)	—	833.4		
La$_{0.5}$Sr$_{0.5}$Fe$_{0.8}$Ga$_{0.2}$O$_{3-\delta}$ (pellet)	834.5	833.2		
La$_{0.5}$Sr$_{0.5}$Fe$_{0.8}$Cr$_{0.2}$O$_{3-\delta}$ (pellet)	835.0	833.0		

All values are in units of electronvolt. Peaks are arbitrarily grouped as La(i) and (ii) groups.

Schrieffer [BCS] theory, which applies to type I superconductors, does not describe superconductivity above 30 K). This problem, along with the reason for the metallic-like conductivity noted above the superconducting temperature, has been recognized since the first successful synthesis of YBa$_2$Cu$_3$O$_{7-\delta}$ (YBCO) in 1986. Note: YBCO now represents the most heavily studied type II superconductor.

Although YBCO is also a perovskite-type oxide, it contains an additional cation within each unit cell. The stoichiometry for YBCO is typically described as AB$_2$C$_3$O$_{7-\delta}$, where A, B, C, and O represent the Ba, Y, Cu, and O ions, and δ the O deficiency (this takes some value between 0 and 1). Like BaTiO$_3$, structural distortions are noted. These occur below 773 K and depend on the value δ. When δ lies between 0 and ~0.7, an orthorhombic structure occurs. The resulting unit cell is shown in Figure 5.20. As can be seen, Cu occupies two different sites and O occupies four different sites. Those in Cu(1) and O(1) sites form

TABLE 6.7   Assignments of the Various C-1s, O-1s, Co-2p$_{3/2}$, Cr-2p$_{3/2}$, Fe-2p$_{3/2}$, Ga-2p$_{3/2}$, Sr-3d$_{5/2}$, and La-3d$_{5/2}$ Peaks for the Mixed Metal Oxides Analyzed

Peak	B.E.$_{XPS}$ (eV)	Assignment
C-1s(I)	284.8	Adventitious C
C-1s(II)	288.5	Surface CO_3^{2-}
O-1s(I)	528.5	Bulk O^{2-}
O-1s(II)	529.8	Surface Sr–O
O-1s(III)	531.0	§Surface B–O
O-1s(IV)	532.0	Surface CO_3^{2-}
O-1s(V)	533.2	§Surface O–O
Co-2p3/2	780.4	Co^{3+}
Cr-2p3/2	576.8	Cr^{3+}
Fe-2p3/2	710.5	Fe^{3+}
Ga-2p3/2(I)	1116.8	Bulk Ga^{3+}
Ga-2p3/2(II)	1117.9	Surface Ga^{3+}
Sr-3d5/2(I)	131.6	Bulk Sr^{2+}
Sr-3d5/2(II)	133.0	Surface Sr–O
Sr-3d5/2(III)	134.0	Surface Sr–CO_3
La-3d5/2(I)	833.5	Bulk La^{2+}
La-3d5/2(II)	835	¢La–C

Numerals refer to arbitrarily defined groups listed in Tables 6.4–6.6.
Keys to Tables 6.4–6.7
*   Data adjusted to place CO_3^{2-} at 290 eV since the 285 eV C-1s signal was very weak.
** Data adjusted to place C at 286 eV (identified on Sr metal and places E_F at 0 eV).
–   Not observed but may be present.
§   These assignments are tentative (see text).
¢   The carbon speciation could not be identified.

one-dimensional chains, while those in Cu(2), O(2), and O(3) sites form two-dimensional planes. The distortion results from an out-of-plane movement of Cu in the Cu(2) sites and the O in the O(2) and O(3) sites. When δ lies between ~0.7 and 1.0, a tetragonal structure is noted. This primarily results from the movement of Ba along the c-axis toward Y. The a, b, and c unit cell parameters equate to 3.820, 3.885, and 11.683 Å for orthorhombic YBCO and 3.857, 3.857, and 11.819 Å for tetragonal YBCO. Below 773 K, reduction of YBCO occurs through the removal of O from O(1) sites. Reduction can be induced through thermal treatment of prolonged UHV exposure.

Orthorhombic YBCO also exhibits highly complex spectra in both XPS and Auger electron spectroscopy (AES). This stems in part from

(a) The complex electronic structure of these surfaces
(b) The difficulty in collecting artifact-free spectra
(c) The structural modifications that can occur during analysis

Note: UHV cleaved surfaces do not necessarily represent the bulk structure. This is realized since cleavage tends to occur along defective planes.

The purpose of this study is to examine core-level photoelectron spectra before, during, and after the reduction of YBCO such that an internally consistent data set could be arrived at. Reduction (removal of O) was carried out through extended UHV exposure along with subsequent thermal treatment (all carried out *in situ*). Some related oxides, carbonates, and sulfates, namely, CuO, Cu_2O, $CuCO_3$, $CuSO_4$, $BaCO_3$, $CaTiO_3$, $SrTiO_3$, and $BaTiO_3$ were also examined.

The CuO, Cu_2O, $CuCO_3$, $CuSO_4$, and $BaCO_3$ powders were obtained from Aldrich (>99.9%). The $CaTiO_3$, $SrTiO_3$, and $BaTiO_3$ consisted of $5 \times 5 \times 1$ mm polished single-crystal wafers obtained from MTI and Escete BV. The YBCO samples consisted of *c*-axis-oriented 200-Å films deposited on MgO (Seo, 2005). Deposition was conducted via reactive thermal coevaporation of Cu, Y, and Ba atomic species under an O_2 pressure of 10^{-2} to 10^{-4} Torr with the substrate temperature held at 950 K. Subsequent annealing was carried out at 770 K under an O_2 pressure of 100–300 Torr to maximize O uptake. Resistivity measurements revealed metallic conductivity at 300 K and superconductivity below 92 K for the freshly prepared films (Seo, 2005). X-ray diffraction (XRD) analysis at 300 K revealed an orthorhombic structure (*c* parameter of 11.70 Å). These films were cut into 5×5 mm pieces, and then wet etched with 0.1% Br–ethanol solution just prior to insertion into the XPS instrument. Partial reduction of these films was carried *in situ* through extended UHV exposure (2×10^{-9} Torr). Complete reduction occurred on heating to 750 K in UHV. Subsequent XRD analysis carried out at 300 K revealed the tetragonal phase (*c* parameter is 11.86 Å).

Representative spectra obtained over the O-1s, Cu-2p, valence, and the Cu-*LMM* regions are shown in Figure 6.9.

These were collected from

(a) The as-received orthorhombic YBCO film
(b) The same YBCO film following extended UHV exposure (this will cause partial reduction due to the removal of O from most outer layers)
(c) The same YBCO film following thermal treatment *in vacuo* at 775 K (this induces a complete reduction of the YBCO film resulting in the phase change)

The O-1s spectra recorded from orthorhombic YBCO appears consistent with previous studies (Vasquez 91, 92, 94) in that two broad

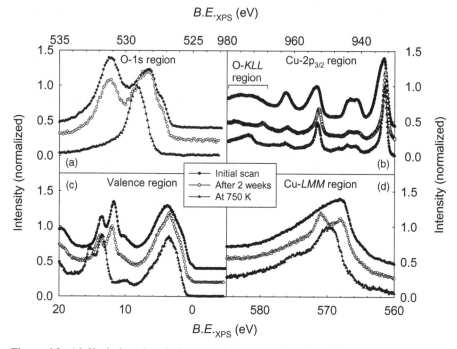

Figure 6.9. Al-$K\alpha$-induced emissions over (a) the O-1s region, (b) the Cu-2p$_{3/2}$ region, (c) the valence region, and (d) the Cu-$L_3M_{45}M_{34}$ region recorded from an in initially orthorhombic YBCO thin film. As noted, these were collected at 300 K as a function of UHV exposure time and at 750 K. All spectra were collected at a takeoff angle of 70° and were normalized to unity. These are then shifted along the intensity axis to retain clarity.

features are noted, one at ~528.2 eV and the other at ~531.7 eV. Due to the complexity of the spectra, curve fitting was applied as detailed in Figure 5.20. In short, the former was fitted using contributions centered at 527.1, 528.2, and 528.7 eV. These have been assigned to O_2^- in (1) bulk chain sites (O(1) sites), (2) bulk plane sites (O(2) and O(3) sites), and (3) surface oxide sites, respectively. The peak at ~531.7 eV is simply assigned to O present in surface contaminants. All assignments and references are listed in Table 6.8.

Angle-resolved studies carried out prior to and following extended UHV exposure allow the reduction experienced at the outer surface to be examined. For example, spectra from the Cu-2p region do reveal a loss of the satellite structure signifying the conversion of Cu^{2+} to Cu^+. More interesting, however, are the spectral variations noted over the Cu-$L_3M_{45}M_{34}$ region since these reveal peaks that can be used to follow

TABLE 6.8 O-1s $B.E._{XPS}$ Values as Derived from Spectra Collected from YBCO and Various Related Copper Oxides and Carbonates at 300 K, as Collected by the Author (van der Heide, 2005)

YBCO	O-1s	Assignment	Literature
C-1s	284.6	Adventitious C	284.6 (Vas 92, 94, B&S)
O-1s	527.4	O^{2-} in chain(b) in YBCO(o)	527.1 (Vas 92)
O-1s	528.2	O^{2-} in plane(b) in YBCO(o)	528.0 (Vas 92)
O-1s	529.2*	O^{2-} in Cu^{2+}–O(s) on YBCO(o)	528.7 (Vas 92)
O-1s	529.8*	O^{2-} in Cu^{+}–O(s) on YBCO(o) ?	—
O-1s	531.7–531.9*	O^{2-} in C=O and C–OH on YBCO(o)	532.3 (Vas 91)
O-1s	532.3*	O in O_2(s) on YBCO(o)	533.3 (Tet)
O-1s	529.2	O^{2-} in plane(b) in YBCO(t)	528.8 (Vas 91)
Other matrices			
O-1s	530.3	O^{2-} in Cu_2O	530.4 (B&S)
O-1s	529.7	O^{2-} in CuO	529.6 (B&S)
O-1s	531.8	O^{2-} in $CuCO_3$	531.5 (NIST)
O-1s	531.8	O^{2-} in $BaCO_3$	532.2 (NIST)
O-1s	532.0	O^{2-} in $CuSO_4$	532.2 (NIST)

Curve fitting of the form shown in Figure 5.20 was used in defining these $B.E._{XPS}$ values. YBCO(o) refers to the orthorhombic phase, while YBCO(t) refers to the tetragonal phase. All values are in units of electronvolt with respect to E_F.

Reference key
Vas 91 Data from Vasquez et al. (1991).
Vas 92 Data from Vasquez et al. (1992).
Vas 94 Data from Vasquez (1994).
B&S   Data from Briggs and Seah (1990).
NIST   Data from Wagner et al. (2003).
Tet   Data from Teterin et al. (1998).
* Surface-bound species.

the reduction occurring over the surface layers. The above variations are shown in Figure 6.10.

The Cu spectra are of particular interest since Cu is the only cation in YBCO that can exist in multiple oxidation states (Ba exists as Ba^{2+} and Y as Y^{3+}). Indeed, there have been suggestions that Cu^{3+} exists in YBCO (Herzog et al., 1988). As discussed below, this is incorrectly understood on the basis of charge neutrality arguments, which imply that either Cu^{3+} or O^- must exist when δ decreases below 0.7 in YBCO. YBCO with δ less than 0.7 is, however, not ionic (this displays metallic properties at room temperature). Orthorhombic YBCO also displays antiferromagnetic ordering arising from superexchange interactions. Such interactions stem from the preference of the spin of an unpaired valence electron in a magnetic ion to align with the spin of an electron

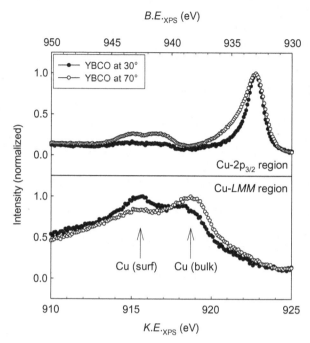

Figure 6.10. X-ray-induced (a) photoelectron emissions over the Cu-2p region and (b) Auger electron emissions over the Cu-$L_3M_{45}M_{34}$ region. These are recorded at 30° and 70° takeoff angles from the orthorhombic YBCO thin film exposed to UHV for the listed times at 300 K. All spectra are normalized to unity.

from a neighboring ligand present in the hybrid orbital formed. In the case of tetragonal YBCO, this interaction occurs between the unpaired electron in the Cu^{2+} d$x^2 - y^2$ level and electrons in the O-2px or 2py levels. Aside from the ordering induced, these electrons experience greater delocalization than would otherwise be expected. The charge discrepancy inferred by charge neutrality arguments is thus explained in terms of hole doping, with the hole (a positive charge) residing in what are referred at as *Zhang–Rice singlet* states. Such states describe a highly mobile hole dispersed among four O^{2-} ions surrounding a Cu^{2+} ion as illustrated in Figure 6.11 (Zhang and Rice, 1988).

The Cu^{2+} ions in YBCO are thus believed to exist in the $d^9 L^{-1}$ state as opposed to the $d^9 L$ state for Cu^{2+} in CuO. The satellites observed in the Cu-2p spectra, which incidentally confirms the presence of Cu^{2+}, must therefore result from the $c^{-1} d^9 L^{-1} \rightarrow c^{-1} d^{10} L^{-2}$ transition (see Section 5.1.1.3.2.1). The initial reduction would then result in the loss of doped holes. This would leave the Cu^{2+} in the d^9 state. Further reduction, which also induces the phase transition from orthorhombic YBCO

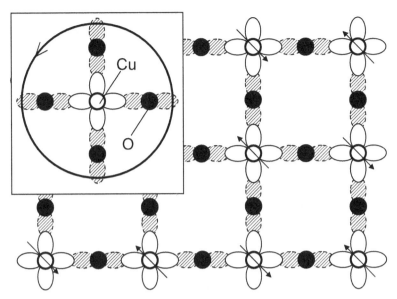

Figure 6.11. A schematic illustration of the superexchange interactions that occur between neighboring Cu^{2+} (core represented by white circles) and O^{2-} (core represented by black circles) in the CuO plane (along the a–b plane shown in Fig. 5.19) when $\delta < 0.7$ for orthorhombic YBCO. The lobes depict the Cu-3d$x^2 - y^2$ (white) and O-2px orbitals (gray). This results in the antiferromagnetic lattice in this plane as indicated by the diagonal arrows. In the inset is depicted a Zhang–Rice singlet state with the doped hole believed to reside among four O ions (the orbit is represented by the large solid circle).

to tetragonal YBCO, would then produce Cu^+ in the d^{10} state. The lack of Cu-2p satellites supports this assignment. Peak assignments based on these arguments are listed in Table 6.9. This includes the Auger parameters (also plotted in Figure 5.19).

The assignments of the photoelectron peaks from the remaining cations (Y and Ba) are listed in Table 6.10 along with literature values. As can be seen, good agreement is noted in all cases.

6.2 SUMMARY

This chapter illustrates several examples of how XPS can be applied in highly specific research projects. Whether examining the capabilities of XPS itself or examining the physical properties of new and/or novel materials, the information derived is a culmination of the knowledge accrued through

TABLE 6.9 Cu $2p_{3/2}$ $B.E._{XPS}$, Cu-$L_3M_{4,5}M_{4,5}$ (Cu L_3VV) $K.E._{XPS}$ and Auger Parameter (A.P) Values from Copper Oxides, Carbonates, and Sulfates along with YBCO at the Listed Temperatures

Element (State) and Temperature	Cu-$2p_{3/2}$	(Literature Values)	Cu L_3VV	(Literature Values)	A.P
Cu^0 (d^9s^2 in Cu) at 300 K	932.7	(932.7 [B&S])	916.7	(916.65 [B&S])	1849.4
Cu^+ (d^{10} in Cu_2O) at 300 K	932.4	(932.4 [B&S])	916.9	(916.8 [B&S])	1849.3
Cu^{2+} (d^9 in CuO) at 300 K	933.7	(933.8 [B&S])	917.8	(917.9 [B&S])	1851.5
Cu^{2+} (d^9 in $CuCO_3$) at 300 K	935.0	(935.0 [NIST])	916.5	(916.3 [NIST])	1851.5
Cu^{2+} (d^9 in $CuSO_4$) at 300 K	935.3	(935.5 [NIST])	916.0	(915.6 [NIST])	1851.3
Cu^{2+} (d^9 in CuF_2) at 300 K	936.8	(936.8 [NIST])	915.2	(914.8 [NIST])	1852.0
Cu^+ (d^{10} in YBCO(t)) at 750 K	933.1	(932.2 [Her])	917.0	(–)	1849.4
Cu^+ (d^{10} in YBCO(o)) at 300 K	~933	(932.2 [Her])	915.5	(–)	1848.5
Cu^{2+} (d^9 in YBCO(o)) at 300 K	932.9	(933.6 [Her])	918.7	(–)	1851.6
Cu^{2+} (d^9L^{-1} in YBCO(o)) at 300 K	934.3	(934.1 [Her])	917.1	(–)	1851.4

YBCO(o) refers to the orthorhombic phase, while YBCO(t) refers to the tetragonal phase. All values are in units of electronvolt with respect to E_F.

TABLE 6.10 Y and Ba $B.E._{XPS}$ Values Recorded from Various Copper Oxides and Carbonates along with YBCO in Its Various States at the Listed Temperatures Are Tabulated Below

YBCO	$B.E._{XPS}$ (eV)	Assignments	Temperature (K)	Literature Values
Y-3d$_{5/2}$	155.7	Y^{3+} in YBCO(o)	At 300	155.8 (Vas 91)
	156.5	Y^{3+} in YBCO(t)	At 750	156.4 (Vas 91)
	157.4*	YO (contaminants)	At 300	—
Ba-3d$_{5/2}$	777.8	Ba^{2+} in YBCO(o)	At 300	777.6 (Vas 91), 778.0 (Nag)
	779.5	Ba^{2+} in YBCO(t)	At 750	778.6 (Vas 91), 779.3 (Nag)
	780.2*	BaO (contaminants)	At 30	—
Ba-4d$_{5/2}$	87.0	Ba^{2+} in YBCO(o)	At 300	87.2 (Vas 91)
	88.7	Ba^{2+} in YBCO(t)	At 750	88.2 (Vas 91)
	89.4*	BaO (contaminants)	At 300	—
Ba-5p$_{1/2}$	11.8	Ba^{2+} in YBCO(o)	At 300	12.0 (Vas 91)
	13.5	Ba^{2+} in YBCO(t)	At 750	12.9 (Vas 91)
	14.5*	BaO (contaminants)	At 300	—

YBCO(o) refers to the orthorhombic phase, while YBCO(t) refers to the tetragonal phase. All values are in units of electronvolt with respect to E_F.

Key for Tables 6.9 and 6.10

Vas 91   Data from Vasquez et al. (1991).
Vas 92   Data from Vasquez et al. (1992).
Vas 94   Data from Vasquez (1994).
B&S   Data from Briggs and Seah (1990).
NIST   Data from Wagner (2003).
Tet   Data from Teterin et al. (1998).
Nag   Data from Nagoshi et al. (1995).
Her   Data from Herzog et al. (1988).

(a) Understanding the importance of surfaces, surface analysis, surface science, and in particular, XPS. This is covered in Chapter 1.

(b) Understanding the electronic structure of atoms, ions, molecules, and solids, and the nomenclature used. This is covered in Chapter 2.

(c) Understanding the instrumentation requirements needed in collected XPS spectra and how these influence the data. This is covered in Chapter 3.

(d) Understanding the factors that dictate the intensities of the photoelectron emission collected and how these are quantified. This is covered in Chapter 4.

(e) Understanding the factors describing photoelectron *B.E.* shifts, along with the introduction of additional features in XPS spectra, and how these can be useful in speciation identification. This is covered in Chapter 5.

With this knowledge, XPS has and will continue to play a major role in surface analysis. Indeed, it is difficult to imagine life today without the many technological breakthroughs we have become accustomed to, some of which owe a lot to surface science (if you cannot see what you have, you cannot predict on how to improve upon it).

APPENDICES

APPENDIX A

PERIODIC TABLE OF THE ELEMENTS

X-ray Photoelectron Spectroscopy: An Introduction to Principles and Practices,
First Edition. Paul van der Heide.
© 2012 John Wiley & Sons, Inc. Published 2012 by John Wiley & Sons, Inc.

Periodic Table

1	2	3	4	5	6	7	8	9	10	11	12	13	14	15	16	17	18
hydrogen 1 **H** 1.0079																	helium 2 **He** 4.0026
lithium 3 **Li** 6.941	beryllium 4 **Be** 9.0122											boron 5 **B** 10.811	carbon 6 **C** 12.01	nitrogen 7 **N** 14.007	oxygen 8 **O** 15.999	fluorine 9 **F** 18.998	neon 10 **Ne** 20.180
sodium 11 **Na** 22.990	magnesium 12 **Mg** 24.305											aluminium 13 **Al** 26.982	silicon 14 **Si** 28.086	phosphorus 15 **P** 30.974	sulfur 16 **S** 32.065	chlorine 17 **Cl** 35.453	argon 18 **Ar** 39.948
potassium 19 **K** 39.098	calcium 20 **Ca** 40.078	scandium 21 **Sc** 44.956	titanium 22 **Ti** 47.867	vanadium 23 **V** 50.942	chromium 24 **Cr** 51.996	manganese 25 **Mn** 54.938	iron 26 **Fe** 55.845	cobalt 27 **Co** 58.933	nickel 28 **Ni** 58.693	copper 29 **Cu** 63.546	zinc 30 **Zn** 65.39	gallium 31 **Ga** 69.723	germanium 32 **Ge** 72.61	arsenic 33 **As** 74.922	selenium 34 **Se** 78.96	bromine 35 **Br** 79.904	krypton 36 **Kr** 83.80
rubidium 37 **Rb** 85.468	strontium 38 **Sr** 87.62	yttrium 39 **Y** 88.906	zirconium 40 **Zr** 91.224	niobium 41 **Nb** 92.906	molybdenum 42 **Mo** 95.94	technetium 43 **Tc** [98]	ruthenium 44 **Ru** 101.07	rhodium 45 **Rh** 102.91	palladium 46 **Pd** 106.42	silver 47 **Ag** 107.87	cadmium 48 **Cd** 112.41	indium 49 **In** 114.82	tin 50 **Sn** 118.71	antimony 51 **Sb** 121.76	tellurium 52 **Te** 127.60	iodine 53 **I** 126.90	xenon 54 **Xe** 131.29
caesium 55 **Cs** 132.91	barium 56 **Ba** 137.33	57–70 *	lutetium 71 **Lu** 174.97	hafnium 72 **Hf** 178.49	tantalum 73 **Ta** 180.95	tungsten 74 **W** 183.84	rhenium 75 **Re** 186.21	osmium 76 **Os** 190.23	iridium 77 **Ir** 192.22	platinum 78 **Pt** 195.08	gold 79 **Au** 196.97	mercury 80 **Hg** 200.59	thallium 81 **Tl** 204.38	lead 82 **Pb** 207.2	bismuth 83 **Bi** 208.98	polonium 84 **Po** [209]	astatine 85 **At** [210]	radon 86 **Rn** [222]
francium 87 **Fr** [223]	radium 88 **Ra** [226]	89–102 **	lawrencium 103 **Lr** [262]	rutherfordium 104 **Rf** [261]	dubnium 105 **Db** [262]	seaborgium 106 **Sg** [266]	bohrium 107 **Bh** [264]	hassium 108 **Hs** [269]	meitnerium 109 **Mt** [268]	ununnilium 110 **Uun** [271]	unununium 111 **Uuu** [272]	ununbium 112 **Uub** [277]		ununquadium 114 **Uuq** [289]			

* Lanthanide series

lanthanum 57 **La** 138.91	cerium 58 **Ce** 140.12	praseodymium 59 **Pr** 140.91	neodymium 60 **Nd** 144.24	promethium 61 **Pm** [145]	samarium 62 **Sm** 150.36	europium 63 **Eu** 151.96	gadolinium 64 **Gd** 157.25	terbium 65 **Tb** 158.93	dysprosium 66 **Dy** 162.50	holmium 67 **Ho** 164.93	erbium 68 **Er** 167.26	thulium 69 **Tm** 168.93	ytterbium 70 **Yb** 173.04

** Actinide series

actinium 89 **Ac** [227]	thorium 90 **Th** 232.04	protactinium 91 **Pa** 231.04	uranium 92 **U** 238.03	neptunium 93 **Np** [237]	plutonium 94 **Pu** [244]	americium 95 **Am** [243]	curium 96 **Cm** [247]	berkelium 97 **Bk** [247]	californium 98 **Cf** [251]	einsteinium 99 **Es** [252]	fermium 100 **Fm** [257]	mendelevium 101 **Md** [258]	nobelium 102 **No** [259]

Figure A.1. The periodic table of the elements based on atomic weight and reactivity. The elements are represented by their chemical symbol. The atomic number (Z) and the atomic weight are listed above and below the respective chemical symbol. *Note:* Elements with bracketed atomic weights do not occur naturally, and hence are not included in the list of binding energies or the list of detectable elements (to be precise, these are detectable).

APPENDIX B

BINDING ENERGIES ($B.E._{XPS}$ OR $B.E._{XRF}$) OF THE ELEMENTS

B.1   1S-3S, 2P-3P, AND 3D VALUES

X-ray photoelectron spectroscopy (XPS), ultraviolet photoelectron spectroscopy (UPS), or X-ray fluorescence (XRF) core-level values were recorded for the elements in their natural form (N, O, F, and Cl are from ionic solids) in units of electronvolt. Note: Both XPS and XRF suffer similar core hole-induced polarization effects. Values for elements/molecules in the gas phase are referenced to E_{vac}, while those in the solid phase are referenced to E_F. Values with the exception of those marked with *, [†], [a], or [b] were taken from Bearden and Burr (1967). Those marked with * are from Cardona and Ley (1978), while those marked with [†] are from Fuggle and Mårtensson (1980). Those marked with [a] were further modified by Williams (2001), while those marked with [b] were added from Hüfner (2003). Note: Speciation can induce variations in these values as illustrated in Figure 5.7 for the O-1s photoelectron emissions.

X-ray Photoelectron Spectroscopy: An Introduction to Principles and Practices, First Edition. Paul van der Heide.
© 2012 John Wiley & Sons, Inc. Published 2012 by John Wiley & Sons, Inc.

Element	1s	2s	$2p_{1/2}$	$2p_{3/2}$	3s	$3p_{1/2}$	$3p_{3/2}$	$3d_{3/2}$	$3d_{5/2}$
H (H$_2$ gas)	13.6								
He (He gas)	24.6*								
Li	54.7*								
Be	111.5*								
B	188*								
C (graphite)	284.7[b]								
N (N$_2$ gas)	409.9*	37.3*							
N (ionic solid)	399.0[b]	12.0[b]							
O (O$_2$ gas)	543.1*	41.6*							
O (ionic solid)	531.0[b]	22.0[b]							
F (F$_2$ gas)	696.7*	—							
F (ionic solid)	686.0[b]	31.0[b]							
Ne (Ne gas)	870.2*	48.5*	21.7*	21.6*					
Na	1070.8[†]	63.5[†]	30.81[a]	30.65[a]					
Mg	1303.0[†]	88.7	49.78	49.50					
Al	1559.6	117.8	72.95	72.55					
Si	1839	149.7*[a]	99.82	99.42					
P	2145.5	189*	136*	135*					
S	2472	230.9	163.6*	162.5*					
Cl (Cl$_2$ gas)	2822.4	270*	202*	201*					
Cl (ionic solid)	—	270[b]	202[b]	200[b]					
Ar (Ar gas)	3205.9*	326.3*	250.6[†]	248.4*	29.3*	15.9*	15.7*		
K	3608.4*	378.6*	297.3*	294.6*	34.8*	18.3*	18.3*		
Ca	4038.5*	438.4[†]	349.7[†]	346.2[†]	44.3[†]	25.4[†]	25.4[†]		
Sc	4492	498.0*	403.6*	398.7*	51.1*	28.3*	28.3*		
Ti	4,966	560.9[†]	460.2[†]	453.8[†]	58.7[†]	32.6[†]	32.6[†]		
V	5,465	626.7[†]	519.8[†]	512.1[†]	66.3[†]	37.2[†]	37.2[†]		
Cr	5989	696.0[†]	583.8[†]	574.1[†]	74.1[†]	42.2[†]	42.2[†]		
Mn	6539	769.1[†]	649.4[†]	638.7[†]	82.3[†]	47.2[†]	47.2[†]		

Fe	7112	844.6†	719.9†	706.8†	91.3†	52.7†	52.7†		
Co	7709	925.1†	793.2†	778.1†	101.0†	58.9†	58.9†tb		
Ni	8333	1008.6†	870.0†	852.7†	110.8†	68.0†	67.2†		
Cu	8979	1096.7†	952.3†	932.7†	122.5†	77.3†	75.1†		
Zn	9659	1196.2*	1044.9*	1021.8*	139.8*	91.4*	89.6*		
Ga	10,367	1299.0*a	1143.2†	1116.4†	159.5†	103.5†	100.0†	18.7†	18.7†
Ge	11,103	1414.6*a	1248.1*a	1217.0*a	180.1*	124.9*	120.8*	29.8	29.2
As	11,867	1527.0*a	1359.1*a	1323.6*a	204.7*	146.2*	141.2*	41.7*	41.7*
Se	12,658	1652.0*a	1474.3*a	1433.9*a	229.6*	166.5*	160.7*	55.5*	54.6*
Br	13,474	1782*	1596*	1550*	257*	189*	182*	70*	69*
Kr (Kr gas)	14,326	1921	1730.9*	1678.4*	292.8*	222.2*	214.4	95.0*	93.8*
Rb	15,200	2065	1864	1804	326.7*	248.7*	239.1*	113.0*	112*
Sr	16,105	2,216	2,007	1,940	358.7	280.3†	270.0†	136.0†	134.2†
Y	17,038	2373	2156	2080	392.0*a	310.6*	298.8*	157.7†	155.8†
Zr	17,998	2532	2307	2223	430.3†	343.5†	329.8†	181.1†	178.8†
Nb	18,986	2,698	2465	2371	466.6†	376.1†	360.6†	205.0†	202.3†
Mo	20,000	2866	2625	2520	506.3†	411.6†	394.0†	231.1†	227.9†
Tc	Not listed since this does not occur naturally								
Ru	22,117	3224	2967	2838	586.1*	483.5*	461.4†	284.2†	208.0†
Rh	23,220	3412	3146	3004	628.1†	512.3†	496.5†	311.9†	307.2†
Pd	24,350	3604	3330	3173	671.6†	559.9†	532.3†	340.5†	335.2†
Ag	25,514	3806	3524	3351	719†	603.8†	573.0†	374.0†	368.3†
Cd	26,711	4018	3727	3538	772.0†	652.6†	618.4†	411.9†	405.2†
In	27,940	4238	3938	3730	827.2†	703.2†	665.3†	451.4†	443.9†
Sn	29,200	4465	4156	3929	884.7†	756.5†	714.6†	537.5†	528.2†
Sb	30,491	4698	4380	4132	946†	812.7†	766.4†	537.5†	528.2†
Te	31,814	4939	4612	4341	1006†	870.8†	820.0†	582.4†	573.0†
I	33,169	5188	4852	4557	1072*	931*	875*	630.8*	619.3*
Xe (Xe gas)	34,561	5453	5107	4786	1148.7*a	1002.1*	934.6*	689.0*	676.4*
Cs	35,985	5714	5329	5012	1211*a	1071*	1003*	740.5*	726.6*

(Continued)

(Continued)

Element	1s	2s	2p$_{1/2}$	2p$_{3/2}$	3s	3p$_{1/2}$	3p$_{3/2}$	3d$_{3/2}$	3d$_{5/2}$
Ba	37,441	5989	5627	5247	1293*[a]	1137*[a]	1063*[a]	795.7*	780.5*
La	38,925	6266	5891	5483	1362*[a]	1209*[a]	1128*[a]	853*	836*
Ce	40,443	6549	6164	5723	1436*[a]	1274*[a]	1187*[a]	902.4*	883.8*
Pr	41,991	6835	6440	5964	1511	1339	1242	948.3*	928.8*
Nd	43,569	7126	6722	6208	1575	1403	1297	1003.3*	980.4*
Pm	Not listed since this does not occur naturally								
Sm	46,834	7737	7312	6716	1723	1541	1420	1110.9*	1083.4*
Eu	48,519	8052	7617	6977	1800	1614	1481	1158.6*	1127.5*
Gd	50,239	8376	7930	7243	1881	1688	1544	1221.9*	1189.6*
Tb	51,996	8708	8252	7514	1968	1768	1611	1276.9*	1241.1*
Dy	53,789	9046	8581	7790	2047	1842	1676	1333	1292.6*
Ho	55,618	9394	8918	8071	2128	1923	1741	1392	1351
Er	57,486	9751	9264	8358	2207	2006	1812	1453	1409
Tm	59,390	10,116	9617	8648	2307	2090	1885	1515	1468
Yb	61,332	10,486	9978	8944	2398	2173	1950	1576	1528
Lu	63,314	10,870	10,349	9244	2491	2264	2024	1639	1589
Hf	65,351	11,271	10,739	9561	2601	2365	2108	1716	1662
Ta	67,416	11,682	11,136	9881	2708	2469	2194	1793	1735
W	69,525	12,100	11,544	10,207	2820	2575	2281	1872	1809
Re	71,676	12,527	11,959	10,535	2932	2682	2367	1949	1883
Os	73,871	12,968	12,385	10,871	3049	2792	2457	2031	1960
Ir	76,111	13,419	12,824	11,215	3174	2909	2551	2116	2040
Pt	78,395	13,880	13,273	11,564	3296	3027	2645	2202	2122
Au	80,725	14,353	13,734	11,919	3425	3148	2743	2291	2206
Hg	83,102	14,839	14,209	12,284	3562	3276	2847	2385	2295

B.2 4S-5S, 4P-5P, AND 4D VALUES

XPS or XRF core-level values recorded for the elements in their natural form (N, O, F, and Cl are from ionic solids) in units of electronvolt. Note: Both suffer similar core hole-induced polarization effects. Values for the rare gases are referenced to E_{vac}, while those in the solid phase are referenced to E_F. Values with the exception of those marked with *, [†], or [a] were taken from Bearden and Burr (1967). Those marked with * are from Cardona and Ley (1978), while those marked with [†] are from Fuggle and Mårtensson (1980). Those marked with [a] were further modified by Williams (2001).

Element	4s	$4p_{1/2}$	$4p_{3/2}$	$4d_{3/2}$	$4d_{5/2}$	5s	$5p_{1/2}$	5p3/2
Kr (Kr gas)	27.5*							
Rb	30.5*	16.3*	15.3*					
Sr	38.9[†]	21.3	20.1[†]					
Y	43.8*	24.4*	23.1*					
Zr	50.6[†]	28.5[†]	27.1[†]					
Nb	56.4[†]	32.6[†]	30.8[†]					
Mo	63.2[†]	37.6[†]	35.3[†]					
Tc	Not listed since this does not occur naturally							
Ru	75.0[†]	46.3[†]	43.2[†]					
Rh	81.4*[a]	50.5[†]	47.3[†]					
Pd	87.1*[a]	55.7[†a]	50.9[†]					
Ag	97.0[†]	63.7[†]	58.3[†]					
Cd	109.8[†]	63.9[†a]	63.9[†a]					
In	122.9[†]	73.5[†a]	73.5[†a]	17.7[†]	16.9[†]			
Sn	137.1[†]	83.6[†a]	83.6[†a]	24.9[†]	23.9[†]			
Sb	165.2[†]	95.6[†a]	95.6[†a]	33.3[†]	32.1[†]			
Te	169.4[†]	103.3[†a]	103.3[†a]	41.9[†]	40.4[†]			
I	186*	123*	123*	50.6	48.9			
Xe (Xe gas)	213.2*	146.7	145.5*	69.5*	67.5*	23.3*		
Cs	232.3*	173.4*	161.3*	79.8*	77.5*	22.7		
Ba	253.5[†]	192	178.6[†]	92.6[†]	89.9[†]	30.3[†]	17.0[†]	
La	274.7*	205.8	196.0*	105.3*	102.5*	34.3*	19.3*	16.8*
Ce	291.0*	223.2	206.5*	109*	—	37.8	19.8*	17.0*
Pr	304.5	236.3	217.6	115.1*	115.1*	37.4	22.3	22.3
Nd	319.2*	243.3	224.6	120.5*	120.5*	37.5	21.1	21.1
Pm	Not listed since this does not occur naturally							
Sm	247.2*	265.6	247.4	129	129	37.4	21.3	21.3
Eu	360	284	257	133	127.7*	32	22	22
Gd	278.6*	286	271	—	142.6*	36	28	21
Tb	396.0*	322.4*	284.1*	150.5*	150.5*	45.6*	28.7*	22.8*
Dy	414.2*	333.5*	293.2*	153.6*	153.6*	49.9*	26.3	26.3
Ho	432.4*	345.5	308.2*	160*	160*	49.3*	30.8*	24.1*

(*Continued*)

(*Continued*)

Element	4s	4p$_{1/2}$	4p$_{3/2}$	4d$_{3/2}$	4d$_{5/2}$	5s	5p$_{1/2}$	5p3/2
Er	449.8*	366.2	320.2*	167.6*	167.6*	50.6*	31.4*	24.7*
Tm	470.9*	385.9*	332.6*	175.5*	175.5*	54.7*	31.8*	25.0*
Yb	480.5*	388.7*	339.7*	191.5*	182.4*	52.0*	30.3*	24.1*
Lu	506.8*	412.4*	359.2*	206.1*	196.3*	57.3*	33.6*	26.7*
Hf	538*	438.2[†]	380.7[†]	220.0[†]	211.5[†]	64.2[†]	38*	29.9[†]
Ta	563.4[†]	463.4[†]	400.9[†]	237.9[†]	226.4[†]	69.7[†]	42.2*	32.7[†]
W	594.1[†]	490.4[†]	423.6[†]	255.9[†]	243.5[†]	75.6[†]	45.3*[a]	36.8[†]
Re	625.4[†]	518.7[†]	446.8[†]	273.9[†]	260.5[†]	83[†]	45.6*	34.6*[a]
Os	658.2[†]	549.1[†]	470.7[†]	293.1[†]	278.5[†]	84*	58*	44.5[†]
Ir	691.1[†]	577.8[†]	495.8[†]	311.9[†]	296.3[†]	95.2*[a]	63.0*[a]	48.0[†]
Pt	725.4[†]	609.1[†]	519.4[†]	331.6[†]	314.6[†]	101.7*[a]	65.3*[a]	51.7[†]
Au	762.1[†]	642.7[†]	546.3[†]	353.2[†]	335.1[†]	107.2*[a]	74.2[†]	57.2[†]
Hg	802.2[†]	680.2[†]	576.6[†]	378.2[†]	358.8[†]	127[†]	83.1[†]	64.5[†]

APPENDIX C

SOME QUANTUM MECHANICS CALCULATIONS OF INTEREST

The basis of quantum mechanics is the Schrödinger equation. This is generally expressed as

$$H\Psi = E\Psi,$$

where E refers to the energy, Ψ is the wave function (a mathematical description of the motion of an electron), and H is termed the Hamiltonian operator (this can represent any mathematical operator that will yield a finite E when applied to a particular Ψ).

Since quantum mechanics involves the construction of systems (atoms, molecules, and clusters representative of solids) from their base units (electrons and nuclei), they can, in principle, allow for the derivation of literally any physical property of any system. The physical properties range from understanding the electronic structure of static systems to the energetics associated with the interaction experienced by any particle (photon, electron, ion, atom, or molecule) impinging on, or traveling through, a solid (dynamic situations). Since energy is stored within bond vibrations, rotations, translations, and so on, these parameters will by default also yield the thermodynamic parameters of enthalpy, entropy, and Gibbs free energy.

X-ray Photoelectron Spectroscopy: An Introduction to Principles and Practices, First Edition. Paul van der Heide.

Although highly effective, the Schrödinger equation can only be solved exactly from first principles (i.e., from physical constants such as the speed of light) for the hydrogen atom. This stems from the fact that in multielectron systems, the interaction between two or more electrons cannot be accurately derived (a result of the fact that the exact position of an electron at some specific time cannot be defined, i.e., a consequence of wave–particle duality as described by the Heisenberg uncertainty principle).

To get around this problem while simplifying the calculations (these are highly complex, requiring extensive computer time), various approximations are applied. Even the simplest ab initio method, the Hartree–Fock method, takes into account the interaction of electrons with each other by assuming an averaged electron density. More complex versions such as Moller–Plesset (MP2 through MP4), correlation interaction (CI), and variants of them, use *electron correlation* to improve accuracy. Electron correlation describes the interaction of electrons with each other. *Density functional methods* (DFTs) use electron correlations and various experimentally defined constants to provide electron densities rather than wave functions (ab initio methods derive wave functions). Since experimental data are used, DFTs are sometimes referred to as *semiempirical* methods. Semiempirical methods, although also based on the Hartree–Fock formalism, introduces additional experimentally derived parameters to ease the computational burden. The complete neglect of differential overlap (CNDO) is a semiempirical method. Such calculations, however, can only be attempted if good parameter sets exist. Molecular mechanics typically avoids the use of quantum mechanics.

There also exists an array of basis sets. A basis set is a mathematical description of the orbitals the electrons exist within. As would be expected, a more complex basis set provides a more accurate result but is more computationally intensive.

Within the context of XPS, most of the above approaches can provide some approximation of core-level binding energies ($B.E.$s) of any electron within any atom/ion. But it should be noted that without modeling the various final state effects introduced on core hole formation, these values only represent those within an unperturbed atom/ion. Correctly modeling these final state effects poses a major task since numerous processes may be initiated with many introducing additional many-body effects of which little information may be available. For these, DFTs presently appear the more appropriate (these allow core-level electrons to be removed, thereby opening up the possibility of examining many of the subsequent processes introduced while minimizing the expense of such computations). As a result of the complexities involved and the approximations required (introduced to ease the computational burden), such studies should only be carried out by those well versed in the calculations and in collaboration with those experienced in the intricacies of XPS.

SOME STATISTICAL DISTRIBUTIONS OF INTEREST

There are several types of distributions used to model the likelihood (statistics) of particular types of events displaying high frequency and nonzero spread. These exist because the distribution of specific types of events can vary, that is, may be symmetric or nonsymmetric. Those most commonly used in XPS are

(a) Gausian distributions
(b) Poisson distributions
(c) Lorentzian distributions

Gaussian and Poisson distributions are related in that they are extreme forms of the *binomial distribution*. The binomial distribution describes the probability distribution for any number of discrete trials. A Gaussian distribution is therefore used when the probability of an event is large (this results in more symmetric bell-shaped curves), while a Poisson distribution is used when the probability is small (this results in asymmetric curves). The Lorentzian distribution represents the exact solution to the differential equation describing forced resonance; thus, it is the most symmetric of the three.

X-ray Photoelectron Spectroscopy: An Introduction to Principles and Practices,
First Edition. Paul van der Heide.
© 2012 John Wiley & Sons, Inc. Published 2012 by John Wiley & Sons, Inc.

The likelihood of a particular event is *generally* described by the mean, median, and standard deviation (σ). The word "generally" is used since Lorentzian distributions do not have a mean and hence σ values, whereas Gaussian and Poisson distributions do. The median represents the middle of the range of values modeled. Note: A mean is needed to derive σ. The mean, if it exists, represents the average value exhibited by the distribution and σ describes the likelihood from the mean value over which a particular event will occur; that is, 68.3% occur within $\pm 1\sigma$ of the mean value, 95.5% occur within $\pm 2\sigma$, and 99.7% occur within $\pm 3\sigma$ for Gaussian distributions.

D.1 GAUSSIAN DISTRIBUTION

This frequency-type distribution, also called *normal distribution*, is applied when describing events that exhibit a highly symmetric array of possible outcomes (mean and median can differ) with populations falling into the standard deviations listed above. An example of this is the effect an XPS instrument has on the energy of a stream of core-level photoelectrons from a specific level that do not suffer significant final state effects. This is due to the finite width of the energy window over which the electrons are collected. Since an infinitely narrow energy window would actually result in a Lorentzian distribution, the outcome becomes a convolution of the two with the Gaussian form dominating (these are called *Voigt functions*). The full width at half maxima (FWHM) of a Gaussian distribution is commonly used to describe the spread (width). This equates to $2\sqrt{(2 \ln 2)}\sigma$.

D.2 POISSON DISTRIBUTION

This frequency-type distribution is applied when describing events that have a nonsymmetric probability of occurring around some mean value. Typical examples are in particle decay. The event/nonevent is then expressed in units over which the signal decays, that is, e^{-n}, where n is some integer number. In the case of electron path length (described by λ), the event described is energy loss occurring through inelastic collisions. The value of λ is taken as equal to that over which the electron intensity (area) at some specific energy drops by e^{-1} or 61.7% from its original value assuming normal takeoff from a homogeneous solid with no elastic collisions or diffraction effects. A loss of intensity by e^{-2} or 85.6% equates to 2λ and e^{-3}, or 95.7% equates to 3λ.

D.3 LORENTZIAN DISTRIBUTIONS

This frequency-type distribution, also called a *Lorentz distribution* or a *Cauchy distribution* (after Hendrick Lorentz and Augustin Cauchy), is applied when describing events that exhibit a truly symmetric array of possible outcomes as in forced resonance, that is, that due to homogeneous broadening. An example of this is the effect the core hole lifetime has on the energy of the photoelectron emission responsible (an effect resulting from the uncertainty principle as relayed through Fig. 5.1). Note: All core-level photoelectron peaks not suffering significant final state effects exhibit this distribution, at least before being modified by phonons and the instrument used to record them (these introduce a Gaussian component). As a result, these distributions have a narrower FWHM, but wider tails than comparative Gaussian distributions, and do not have a distinct mean value nor σ (the latter is defined from the former).

APPENDIX E

SOME OPTICAL PROPERTIES OF INTEREST

Ion, electron, and photon beams all share similar optical properties; that is, all can be formed into a collimated (parallel) stream of particles moving in the same direction. This is otherwise referred to as a particle beam. This beam can then be deflected and/or focused at some position in space. Such beams can also suffer the same set of deformations, resulting in loss of spatial definition. If images are generated from these beams in the transverse plane (the plane perpendicular to the beam axis), such deformations will result in a loss of spatial resolution. Two such effects that result in loss of spatial resolution include

(a) Chromatic aberrations
(b) Spherical aberrations

In addition, the smallest probe size (diameter in the transverse plane at the focal point) that can be achieved is fundamentally limited by what is referred to as the diffraction limit. This cannot be improved upon using conventional optics. Note: In transmission electron microscopy (TEM), the diffraction limit is smaller than atomic dimensions. This unusual situation precludes the significance of the diffraction limit in TEM.

X-ray Photoelectron Spectroscopy: An Introduction to Principles and Practices,
First Edition. Paul van der Heide.
© 2012 John Wiley & Sons, Inc. Published 2012 by John Wiley & Sons, Inc.

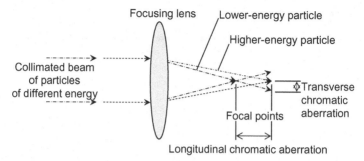

Figure A.2. Pictorial illustration of chromatic aberrations resulting from the energy dispersion in the respective beam in both the longitudinal (parallel to the original beam) and transverse (perpendicular) planes.

E.1 CHROMATIC ABERRATIONS

These arise when the energy of the respective probe particles spans some finite range, that is, a nonmonochromated beam. This is realized since the focal point, and hence the focusing characteristics of any optical element, is a direct function of the probe particle's energy as illustrated in Figure A.2 (energy, wavelength, and frequency are all functions of each other). Note: Some intrinsic spread in energy also remains in monochromatized beams. This fundamental limit exists since, due to the Heisenberg uncertainty principle, a particle's position and momentum (and hence energy) cannot all be absolutely fixed at the same time.

For a beam of fixed current, these effects can be minimized by accelerating the probe beam to higher energies (this minimizes the effective spread in energy). In the case of a charged particle beam, this also reduces space charge effects (coulombic repulsion of charged particles in close proximity) by effectively reducing their proximity to each other. Reducing the beam current density has the same effect for the same reason.

E.2 SPHERICAL ABERRATIONS

These describe the increased focusing/defocusing action experienced by the particles that make up a monochromated beam that are a significant distance from the optical axis as illustrated in Figure A.3.

E.3 DIFFRACTION LIMIT

This describes the fundamental limit of any beam of particles (photons, electrons, atoms, ions, or molecules) passing through an aperture in

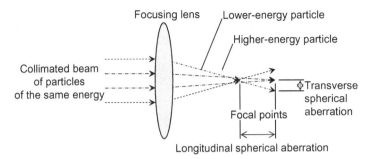

Figure A.3. Pictorial illustration of spherical aberrations resulting from the spatial dispersion in the respective beam in both the longitudinal (parallel to the original beam) and transverse (perpendicular) planes.

conventional systems. This arises from the Heisenberg uncertainty principle, which states that a body's position and momentum cannot both be known at the same time. Thus, the greater the certainty of one of these parameters results in a lesser certainty of the other. Since momentum is a product of mass and velocity, the smaller the mass of the object, the greater the uncertainty in its position. This, in turn, limits the spatial resolution that can be achieved when imaging with lighter particles. This, therefore, explains why

(a) An optical microscope cannot provide images to a spatial resolution better (smaller) than approximately half the wavelength of the photons used in acquiring the image. This results in an ultimate spatial resolution of 200–400 nm in dedicated state-of-the-art microscopes.

(b) A scanning electron microscope operated at 15 keV can provide images with a spatial resolution approaching ~1 nm (electrons have a mass of 9.109×10^{-28} g). A transmission electron microscope can provide subatomic resolution by using higher-energy electrons (up to 300 keV).

(c) A helium ion microscope (similar to a scanning electron microscope but with helium ions used to generate scanning electrons) operated at 45 keV with a point source can provide images with a spatial resolution approaching ~0.25 nm (helium has a mass ~8000 times that of an electron or 3.344×10^{-24} g).

Scanning electron microscopes, helium ion microscopes, and transmission electron microscopes are discussed further in Appendix G.

SOME OTHER SPECTROSCOPIC/ SPECTROMETRIC TECHNIQUES OF INTEREST

There is no *one end all be all* technique, that is, one technique that provides all possible combinations of the information of interest. This stems from the fact that each source/signal combination provides different forms of information, with the technique in question specialized toward maximizing the respective information content contained within. The techniques covered in this section are subdivided according to whether photons, electrons, or ion emissions are recorded. These include

(a) The photon spectroscopies (those that analyze photon emissions) of
- (i) Infrared (IR)-based techniques inclusive of reflection–adsorption infrared spectroscopy (RAIRS), attenuated total reflection (ATR), and diffuse reflectance infrared spectroscopy (DRIFTS)
- (ii) Raman including surface-enhanced Raman spectroscopy (SERS) and tip-enhanced Raman spectroscopy (TERS)
- (iii) Energy-dispersive X-ray analysis (EDX) and wavelength-dispersive X-ray analysis (WDX)
- (iv) X-ray fluorescence (XRF) and total reflection X-ray fluorescence (TXRF)

X-ray Photoelectron Spectroscopy: An Introduction to Principles and Practices,
First Edition. Paul van der Heide.
© 2012 John Wiley & Sons, Inc. Published 2012 by John Wiley & Sons, Inc.

(b) The electron spectroscopies (those that analyze electron emissions) of
  (i)   Ultraviolet photoelectron spectroscopy (UPS)
  (ii) Auger electron spectroscopy (AES)
  (iii) Electron energy loss spectroscopy (EELS), including reflected electron energy loss spectroscopy (REELS) and high-resolution electron energy loss spectroscopy (HREELS)
(c) The ion spectroscopies/spectrometries (those that analyze ion emissions) of
  (i)   Secondary ion mass spectrometry (SIMS)
  (ii) Tomographic atom probe (TAP)
  (iii) The ion scattering methods inclusive of low-energy ion scattering (LEIS), medium-energy ion scattering (MEIS), and Rutherford backscattering (RBS)

The techniques chosen in this section are by no means a complete selection, rather representative in the eyes of the author, of some of the more popular/more interesting surface- or probe-based spectroscopies/ spectrometries, that is, those routinely used to provide chemical information of localized volumes. Microscopies and diffraction techniques used for imaging are discussed in Appendices G and H, respectively.

The original difference between a spectroscopy and spectrometry lies in the fact that in a spectrometry, only one parameter of interest varies (generally, this is intensity), while in a spectroscopy, more than one parameter of intensity can vary (this can be energy, mass, time, intensity, etc.).

XPS spectra comprise intensity versus $B.E._{XPS}$ data. Since both the intensity (the area reveals the concentration) and the $B.E._{XPS}$ (this reveals speciation) of the signal of interest can vary, this technique is defined as a spectroscopy. SIMS spectra comprise intensity versus secondary ion mass data. In this case, only the intensity can vary (secondary ion mass is a constant). As a result, this is defined as a spectrometry. Indeed, ISS in this context should also be defined as ion scattering spectrometry. This definition has, however, not been strictly adhered to over the years; that is, ISS is also commonly referred to as ion scattering spectroscopy.

TAP is included in this section since this has recently experienced extensive commercial development, which should move it from a purely academic methodology to a routinely utilizable technology. Atomic force microscopy (AFM), LEIS, MEIS, and Raman have also experienced extensive development over the last couple of decades. Of the techniques listed, only a few are capable of providing compositional information at the atomic scale. Indeed, TAP can accomplish this in all three dimensions for every isotope of all the elements.

F.1 PHOTON SPECTROSCOPIES

Photon spectroscopies are defined as those techniques that provide chemical information from the sample of interest by sampling the induced photon emissions from the respective sample.

F.1.1 IR, RAIRS, ATR, and DRIFTS

IR-based techniques are those that derive information on the chemical bonding that occurs between bound atoms, whether in the liquid/solid or gas phase. This information is derived by irradiating the sample with photons over the IR range (<1 eV) and sampling the adsorption/transmission that occurs. Adsorption occurs as a result of excitation of valence electrons into higher vibrational states with the energy specific to the masses of the bound atoms and bonding present. The extent of adsorption also relates to the amount of the elements/molecules present. The IR beam can also be focused to allow imaging (diffraction-limited spatial resolution approaching 1 μm). Fourier transform (FT) analysis is typically used since this enhances the data intensity/quality with the acronym *FTIR* used. Fourier transform infrared spectroscopy (FTIR) is, however, not intrinsically surface sensitive since the IR beam passes many micrometers through many solids. IR beams do, however, reflect off of metal layers; thus, by depositing an IR transparent film on a metal layer and recording the reflected signal, some degree of surface specificity can be realized. Although specific sample preparation is required analysis can be carried out under ambient conditions. Greater surface specificity can be realized using IR in the form of RAIRS, ATR, or even DRIFTS.

RAIRS, also called infrared adsorption spectroscopy (IRAS), is a form of FTIR that can provide bonding information as well as concentration if the surface coverage is $>\sim 1 \times 10^{13}$ atoms/cm^2 of any atom/molecule adsorbed on a highly reflective low surface area solid, for example, CO on Pt. This does so by recording the adsorption that occurs upon specular reflection of photons directed at low incidence angles. No prior sample preparation is required.

Although not requiring UHV conditions, this helps in controlling the environment.

ATR is a form of FTIR that can provide bonding information as well as concentration if the surface coverage is $>\sim 1 \times 10^{13}$ atoms/cm^2 of atoms/molecules adsorbed on low surface area solids. In this form, a photon beam is passed through a thin IR transmitting high-density crystal with a high refractive index (ZnSe or Ge) that is in direct contact with the sample of interest. Upon reflection from the crystal face, photon absorption occurs with molecules present at the crystal/sample interface. This occurs since some fraction of the radiation penetrates a small distance into the sample (evanescent wave). The multiple internal reflection (MIR) technique enhances the sensitivity of ATR by bouncing the photons off the surface numerous times. Minimal sample preparation is required and analysis can be carried out under ambient conditions. Research has also been carried out in the area of surface-enhanced infrared spectroscopy (SEIRS) using an ATR-type setup following the development of SERS, the Raman analogue.

DRIFTS is a form of FTIR that can provide bonding information as well as concentration if the surface coverage is $>1 \times 10^{13}$ atoms/cm^2 of atoms/molecules adsorbed on high surface area solids, that is, finely dispersed catalysts. This technique does so by sampling photons experiencing diffuse reflectance. In this mode, some fraction of the photons impinging on a surface is transmitted into the solid. These photons may then be absorbed, further transmitted or reflected out of the solid. The surface-reflected and bulk reemitted components, which when summed, represent the recorded signal. No prior sample preparation is required and analysis can be carried out under ambient conditions.

F.1.2   Raman, SERS, and TERS

Raman-based techniques are those that derive information on the chemical bonding that occurs between bound atoms, whether in the liquid/solid or gas phase. This information is derived by irradiating the sample with photons over the visible/UV range (>1 eV) and by sampling the adsorption/emissions induced. These occur as a result of excitation of valence electrons into higher electronic levels, which then relax back into their original or related vibrational/rotational state. The difference in energy between the original and related states is then measured (this is specific to the masses of the bound atoms and bonding present). The signal intensity also relates to the amount of the elements/molecules present. Furthermore, the laser used can be focused to allow imaging (diffraction-limited spatial resolution to ~1 μm). Raman is closely related to FTIR spectroscopy, with the data being highly complementary. Like FTIR, Raman is not intrinsically surface specific. Raman also suffers from a lack of sensitivity. Recent developments have, however, resulted in the resurgence of Raman in the form of SERS and TERS, both of which are highly region specific.

SERS is a technique capable of providing information from individual molecules. This does so introducing silver or gold nanoparticles since this significantly enhances the Raman signal over highly localized regions. The fact that enhancement factors greater than 10^{10} are realized opens up the possibility of noninvasively examining individual molecules. The significant enhancement is believed to be due to the introduction of a surface plasmon resonance resulting in an electromagnetic enhancement that decays exponentially with distance from the nanoparticle. These can thus be thought of as nanoprobes. Although circumventing the diffraction limit, placement of the particles is random by nature. Minimal sample preparation is required and analysis can be carried out under ambient conditions.

TERS is a technique capable of providing images to a resolution of ~20 nm over prespecified areas on a solid substrate. Although closely related to SERS (TERS relies on the same enhancement provided by the presence of silver), imaging over predefined regions is made possible by supporting a silver or gold nanoparticle on the end of an AFM-like probe tip and by scanning this over the surface of interest, which is also irradiated by photons in the visible/UV region (that needed to generate the Raman signal). Localized Raman signal enhancement is then noted from highly localized regions, that is, those regions within ~10 nm from the tip. This represents another technique that has been able to circumvent the diffraction limit. Minimal sample preparation is required and analysis can be carried out under ambient conditions.

F.1.3 EDX and WDX

EDX and *WDX* are both techniques capable of providing the elemental composition of the outer ~1–5 μm from any solid to a sensitivity of

~0.1% (sensitivity is element dependent). All elements from B to U are detectable. These techniques do so by directing a medium- (5–30 keV if carried out in scanning electron microscopy [SEM] or electron probe microanalysis [EPMA]) to high-energy (up to 300 keV if carried out in TEM) electron beam at the solid of interest at close to normal incidence. This induces core electron emission, along with Auger electrons and/or fluorescence (the latter two stem from the core hole generated by core electron emission). Both EDX and WDX sample the fluorescence (the energy is element specific) but in different fashions. EDX uses either Si(Li) or silicon drift detectors with the latter gaining in popularity. WDX uses a crystal with the angle of reflection based on the impinging photon's energy, that is, by satisfying the Bragg relation. No prior sample preparation is needed, but a minimum of high vacuum (HV) is required.

F.1.4 XRF and TXRF

XRF is a technique capable of providing the elemental composition throughout a specifically derived location with all elements from Na to U detectable with some down to 0.01% and 0.001% (sensitivity is element dependent). It is not a surface-specific technique (the volume probed can extend to 10 μm below the surface) but is covered here since it also identifies the elements based off of the *B.E.*s of the respective electrons (*B.E.*$_{XRF}$). Indeed, much of the higher-energy data in Appendices B and C were derived via XRF. The *B.E.*$_{XRF}$ values are derived in much the same way as in EDX/WDX, the difference being that the initial core hole is generated by an X-ray as opposed to an electron. This induces core electron emission along with Auger electrons and/or fluorescence (the latter two stem from the core hole generated by core electron emission). XRF samples the fluorescence. Neither prior sample preparation nor vacuum conditions are required.

TXRF is a variant of XRF capable of providing the elemental composition of the outer ~1–5 nm from any highly reflective solid to a sensitivity of between 0.01% and 0.001% (sensitivity is element dependent). All elements from S to U are typically detectable (recent developments have extended this range to Na–U). This technique does so by directing a high-energy photon beam (X-rays) at the solid of interest at some small glancing angle. This induces photoelectron emission, along with Auger electrons and fluorescence (the latter two stem from the core hole generated) from the surface region alone. TXRF samples the fluorescence produced (the energies are element specific). No prior sample preparation is needed nor are vacuum conditions required.

F.2 ELECTRON SPECTROSCOPIES

Electron spectroscopies are defined as those techniques that provide chemical information from the sample of interest by sampling the induced electron emissions from the respective sample.

F.2.1 UPS

This technique is capable of providing chemical bonding information, and in very limited cases, elemental composition of the outer ~1–5 nm from any solid. The spatial resolution can be down to ~100 nm. This technique does so by directing a low-energy photon beam (UV energy) at the solid of interest and by recording the photoelectron emissions. The primary difference between UPS and XPS lies in the energy of the photons used; that is, UPS can only sample valence electrons, whereas XPS samples both valence and core electrons. Since the photon energy used in UPS is more closely matched to the valence electron $B.E.$s, this technique is far more sensitive to the valence region than XPS. No prior sample preparation is needed, but UHV is required.

F.2.2 AES

This is a technique capable of providing the elemental composition of the outer ~1–5 nm from any solid, although insulators are difficult, to a sensitivity down to ~0.5% (sensitivity is element dependent) with some speciation information also available. All elements from Li to U are detectable. This technique does so by directing a medium-energy electron beam (5–30 keV) at the solid of interest. This induces valence and core electron emission with the resulting holes producing Auger electrons or fluorescence (only the Auger electrons produced are recorded in AES). Elemental identification is made possible since the Auger energies are element/level specific. Note: Scanning Auger microscopy (SAM) is the scanning form of AES; that is, the highly focused 5- to 30-keV electron beam is scanned across the surface as this allows for the lateral distribution of elements present on the surface to be mapped. Spatial resolution can extend to ~10 nm or better. AES is similar to XPS in that both provide similar information using similar instrumentation. The primary differences lie in the fact that AES provides a far superior spatial resolution, but at the cost of sensitivity (XPS provides slightly better sensitivity). No prior sample preparation is needed, but UHV is required.

F.2.3   EELS, REELS, and HREELS

These are all techniques that directly record the energy loss experienced by an electron beam incident on a solid. The energy loss arises from the excitation induced. This may be in the form of vibrational/vibronic modes as measured by HREELS to core-level excitations as induced by REELS and EELS. As a result, HREELS is also referred to as vibrational energy loss spectroscopy (VELS) and can be considered an electron analogue of Raman spectroscopy. EELS, on other hand, is sometimes referred to as core electron energy loss spectroscopy (CEELS) since core levels are accessed. Since core-level $B.E.$s are element specific, elemental identification of all the elements is possible in principle. Sensitivity is typically in the 1% range but quantification can be difficult due to the high background levels present. These techniques differ primarily in the energy of the incoming electron beam and the instruments these are carried out in. In HREELS, a 1- to 10-eV beam is used in highly specialized HREELS instrumentation. In REELS, a 1- to 25-keV beam is typically used in AES-type instruments. In EELS, higher-energy electrons, that is, up to 300 keV, are typically used in TEM-type instruments. As a result, REELS and HREELS are surface specific to the outer ~2 nm and EELS is volume specific when performed in a TEM. EELS, however, displays the best spatial resolution; that is, this can be down to ~1 nm when performed in a high-resolution TEM instrument. That of REELS can, in principle, reach that of the probe size (this can be as low as ~10 nm in AES instruments). HREELS is not typically used for imaging surfaces. No prior sample preparation is needed (with the exception of EELS if carried out in a TEM instrument), but UHV is required in all cases.

F.3   ION SPECTROSCOPIES/SPECTROMETRIES

Ion spectroscopies are defined as those techniques that provide chemical information from the sample of interest by sampling the induced ion emissions from the respective sample.

F.3.1   SIMS

This technique is split into two subgroups referred to as dynamic SIMS and static SIMS. The *dynamic SIMS* form is capable of providing the elemental composition over any region of any solid to a sensitivity down to ~0.1 ppb (strongly dependent on element and analysis conditions). Both depth and volume maps can be derived to a depth resolution of ~1 nm and a spatial resolution of 100 nm or better (20-nm best reported value). All elements and their isotopes from H to U are detectable. *Static SIMS* displays a decreased sensitivity due to the fact that a small faction of the outermost monolayer only is probed. The reduced damage does, however, allow molecular information to be extracted. Both dynamic and static SIMS attain their information by directing low- to medium-energy (~0.1–25.0 keV) ions (O_2^+, Cs^+, Ar^+, Ga^+, C_{60}^+, Au_{1-3}^+, Bi_{1-3}^+, etc.) at the surface. This induces the removal of surface atoms (sputtering), a small fraction of which departs in the ionized state. The ions are then directed through energy and mass filters (magnetic sector, quadrupole, or time of flight) to allow elemental (isotopic) and/or molecular identification based on their respective mass. No prior sample preparation is needed, but UHV is required.

F.3.2 TAP

This technique, developed from field ionization microscopy (FIM), is capable of providing elemental/isotopic information over any region of any solid to a sensitivity down to ~0.1 ppm. Volume maps can be derived in which the depth and spatial resolution can be <0.3 nm. All elements and their isotopes from H to U are detectable. This technique attains this information by directing a laser at a specifically fabricated tip containing the region of interest (constructed using focused ion beam [FIB] instrumentation). This form of TAP, often called laser-assisted tomographic atom probe (LATAP), induces the removal of atoms, which become ionized by the high potential field surrounding the tip. This field also guides these ions through a time-of-flight mass filter onto a position-sensitive detector. The former allows for elemental identification (isotopic), while the latter allows for the tip volume to be reconstructed at the atomic level. Significant sample preparation (construction of the tip), as well as UHV, is required. This heavily limits the type of samples that can be examined.

F.3.3 Ion Scattering Methods

These comprise several techniques (*LEIS*, *MEIS*, and *RBS*) capable of providing the elemental composition as a function of depth, and in some cases, the lattice structure over the surface region, that is, can extend from <0.3 to 100 nm in depth of any solid. The spatial resolution is typically of the order of 1 mm (recent developments have, however, pushed this into the nanometer range). The elemental sensitivity, which is Z^2 dependent, approaches 0.1% for light elements and 10 ppm for heavier elements. All elements from H to U are detectable depending on the geometry and ions used. These techniques attain this information by analyzing the energy lost by ions (He^{2+}, He^+, Ne^+ Ar^+, or Xe^+ depending on the technique) scattered off the surface of interest. This is possible since the energy loss in monoenergetic ion scattering is specific to the mass of the collision partners (see Equation 4.6) and the depth below the surface at which the collision took place. RBS is the most quantitative of the ion scattering techniques since the high-energy ions used (typically 0.5- to 3.0-MeV He^{2+} ions) only allow interaction between the nuclei of the atoms within the sample of interest and the incoming ions. MEIS uses lower-energy ions (typically 10- to 50-keV He^+, Ne^+, Ar^+, or Xe^+ ions), while LEIS uses even lower-energy ions (typically <10-keV He^+, Ne^+, Ar^+, or Xe^+ ions). These provide improved depth resolution. In the case of LEIS, and MEIS to a lesser extent, the possibility of deriving a surface crystal structure is also realized. As an example, surface atomic positions can be measured to better then 0.01 nm using specialized variants of LEIS, otherwise referred to as time-of-flight scattering and recoiling spectroscopy (TOF-SARS), scattering and recoiling imaging spectroscopy (SARIS), and impact collision ion scattering spectroscopy (ICISS). Quantification, however, becomes increasingly difficult as the scattered ion energy is decreased due to the greater electronic interaction occurring between these ions and the surface probed (this increases ion neutralization probabilities dependent on the surface chemistry). In all cases, little sample preparation is needed, but UHV is required (most important for LEIS).

SOME MICROSCOPIES OF INTEREST

The techniques chosen in this section are representative, in the eyes of the author, of some of the more popular surface- or probe-based microscopies. Note: Microscopies are those used to provide images of localized volumes (little or no chemical information is relayed). Spectroscopies/spectrometries are discussed in Appendix F, while diffraction techniques are discussed in Appendix H. The techniques covered in this section include

(a) Scanning electron microscopy (SEM)

(b) Helium ion microscopy (HIM)

(c) Transmission electron microscopy (TEM)

(d) Scanning probe microscopy (SPM) in the form of

    (i)   Atomic force microscopy (AFM)

    (ii) Scanning near-field optical microscopy (SNOM)

    (iii) Electrical force microscopy (EFM), magnetic force microscopy (MFM), kelvin force microscopy (KFM), conductance atomic force microscopy (CAFM), and tunneling atomic force microscopy (TUNA)

X-ray Photoelectron Spectroscopy: An Introduction to Principles and Practices,
First Edition. Paul van der Heide.
© 2012 John Wiley & Sons, Inc. Published 2012 by John Wiley & Sons, Inc.

(iv) Scanning capacitance microscopy (SCM) and scanning spreading resistance microscopy (SSRM)

(v) Scanning tunneling microscopy (STM)

The developments in this area were spurred, in essence, by the fact that the diffraction limit did not allow for optical microscopes to view regions to a spatial resolution much below 1 μm (see Appendix E). Indeed, the spatial resolution in state-of-the-art optical microscopes is diffraction limited to around 200–400 nm, or half the wavelength of visible light. To improve upon this requires circumventing the diffraction limit (accomplished by the SPM-based techniques) or by reducing the diffraction limit through the use of more energetic probes (TEM uses up to 300-keV electrons to reduce this limit to values well below atomic dimensions) or by using more massive probe particles (HIM uses He ions rather than electrons or photons).

TEM is included in this list since, although not a surface technique, it is capable of providing atomic-scale images of any volume of interest. The many different variants of SPM are included since this is a field that has experienced extensive commercial development over the last two decades.

G.1 SEM

This represents the most heavily used of the microscopies for imaging down to the nanometer scale, a region that is inaccessible to the optical microscopies. Indeed, state-of-the-art FE-SEM instruments, where FE stands for the field emission source used, can display a spatial resolution to ~1 nm. SEM operates by directing a finely focused electron beam (source) of ~15 keV at a solid surface of interest. The intensity of the low-energy secondary electrons generated are then recorded (secondary electrons are recorded since they provide much greater intensities than the Auger or scattered electrons also produced). Distribution maps are then generated by scanning the probe beam over the surface and synchronizing this with the detector. The images generated can then be useful in revealing surface structures, surface topography, and even the dispersion of different matrices. This arises from the fact that different faces, features, and even materials have different secondary electron yields. The spatial resolution is then limited by the probe beam diameter and the interaction volume within the surface. Secondary electrons are generated from depths extending 10 nm or more below the surface (see Fig. 4.6). Since electrons are used, all analysis must be carried out under HV conditions. Analysis at higher pressures can only be carried out in specialized instruments in which the pressure within a localized region around the sample can be increased. No special sample preparation is required.

G.2 HIM

This represents a new technology that uses many of the same principles of SEM. The primary difference is a finely focused He ion beam is used. The end result is a microscopy that can image to a spatial resolution of ~0.25 nm. HIM operates by directing a finely focused He ion beam of energy around 45 keV at the solid surface of interest. The intensity of the low-energy secondary electrons generated is then recorded. Distribution maps are then generated by scanning the probe beam over the surface and synchronizing this with the detector. As with SEM, the images generated can then be useful in revealing surface structures, surface topography, and even the dispersion of different matrices. The spatial resolution is then limited by the probe beam diameter and the interaction volume within the surface. Both are smaller for He ions with respect to energetic electrons. Since ions and electrons are used, all analysis must be carried out under HV conditions. The use of He ions allows for the analysis of insulating samples (can be difficult in SEM). No special sample preparation is required.

G.3 TEM

This represents the most heavily used of the microscopies for imaging down to the atomic scale. Indeed, state-of-the-art TEM instruments, using spherical correctors, can display sub-Angstrom resolution (~0.07 nm). TEM operates by directing a finely focused electron beam (source) of up to 300 keV at a specifically fabricated solid foil (lamella) of less than 100 nm in thickness. The electrons passing through this foil are then detected with the image representative of the atomic positions within the foil (the trajectories of these high-energy electrons passing close to atomic nuclei are deflected according to the number of protons within the respective nuclei). Distribution maps can be generated using a static electron beam (TEM) by scanning the probe beam (scanning transmission electron microscopy [STEM]). The images generated can then be useful in revealing structures and even the dispersion of different matrices. This arises from the fact that different materials have different secondary scattering characteristics. The spatial resolution is primarily limited by the probe beam diameter. Since electrons are used, HV conditions are required. Like TAP, however, a highly specialized and extensive sample preparation is required, with FIB more commonly used. Furthermore, the effects of the ambient environment (noise, etc.) must be minimized.

G.4 SPM (AFM AND STM)-BASED TECHNIQUES

This encompasses a group of techniques (STM, AFM, EFM, CAFM, TUNA, SCAM, SSRP, etc.) that derives nanoscale information by scanning an atomically sharp tip over the outer surface of the substrate of interest and measuring the appropriate response. This response may be in the form of current (I), voltage (V), capacitance (C), force (F), or the distance (D) by which the cantilever supporting the tip is deflected, or a combination of the above. These techniques are not able to directly identify the composition of a material. That said, these can, in special cases, derive the concentration of a particular element if a sufficient amount of information on the sample of interest already exists, that is, can distinguish between an n-type or p-type dopants in Si. With the exception of STM, the advantages of these techniques are that no special sample preparation is required and all analysis can be carried out under ambient conditions. The effects of the ambient environment (noise, electromagnetic fields, etc.) must, however, be minimized.

AFM is capable of routinely providing topographical maps to a spatial resolution of a few nanometers. Samples may be conductors, semiconductors, or insulators. Information is derived by scanning an atomically sharp tip over the surface of interest and by measuring its deflection. Measurement of F versus D curves allows for surface tension and bond strength/elasticity/hardness data to be extracted. AFM allows for topography, mechanical, chemical, and electrical properties to be measured, some simultaneously. Atomic-scale spatial resolution can be achieved, but only in highly specialized instrumentation in which a cooled sample is held under UHV.

SNOM, also referred to as SONM, is a technique capable of providing optical images to a spatial resolution of a few tens of nanometers. This is carried out by passing photons down a capillary tube that is some 20 nm in diameter situated within an AFM-like tip. Images are then derived by bringing the tip to within 10 nm and then scanning this tip/photon source. This near-field approach circumvents the diffraction limit by ensuring all dimensions (capillary diameter and tip-to-surface distance) are much less than the wavelength of the photon source used. As in ATR, this exploits the unique properties of evanescent waves (a near-field standing wave formed at the boundary of different media).

EFM can provide surface potential and charge distribution maps to a spatial resolution of a couple of nanometers. Information is derived by scanning an atomically sharp tip over the surface of interest and by measuring the deflection resulting from the electrostatic force. This is sometimes termed voltage probing.

MFM is similar to EFM except for the fact that magnetic domains are mapped by the tip as opposed to electrostatic domains. This can be done to a spatial resolution of a couple of nanometers.

KFM is closely related to EFM in that it can also provide surface potential maps, but in the form of work function variations. This is derived by measuring the contact potential difference (CPD) between the surface and the tip. This can be done to a spatial resolution of a couple of nanometers.

CAFM, also known as current sensing atomic force microscopy (CSAFM), is capable of providing maps of the electrical conductivity (dI/dV) to a spatial resolution of a couple of nanometers. The sample may be a conductor or semiconductor. Information is derived by measuring the current (100 pA–100 µA) passing to/from the atomically sharp tip and the sample surface of interest. The measurement of I versus F or I versus V curves allows for localized data to be extracted.

TUNA is similar to CAFM in that it is also capable of providing electrical conductivity (dI/dV) maps to a spatial resolution of a couple of nanometers. Samples, however, must be less conductive than those used in CAFM. Information is derived by measuring the current (100 fA–100 pA) passing to/from an atomically sharp tip and the sample surface of interest. Since conductivity is a function of film thickness, leakage paths, charge traps, and so on, these too can be mapped. The measurement of I versus F or I versus V curves allows for localized data to be extracted.

SCM is capable of providing carrier concentration maps to a spatial a resolution of a few tens of nanometers. The sample may be a conductor or a semiconductor. Information is derived by measuring the capacitance variations noted on applying an AC bias between the atomically sharp tip and the sample surface of interest. This is useful for measuring concentrations of active *n*- and *p*-type dopants if present at concentrations greater that ~1e16 atoms/cm^3. Measurement of the *dC/dV* versus *V* curves allows for differentiation between *n*- and *p*-type dopants. The polarization state in ferroelectrics can be derived from the slope of their *C* versus *V* curves.

SSRM, like SCM, can provide carrier concentration maps to a spatial resolution of around 10 nm. The sample may be a conductor or a semiconductor. Information is, however, derived by measuring the electrical conductivity or resistivity of a sample sandwiched between an atomically sharp tip and a conductive back plate. This is useful for measuring concentrations of active *n*- and *p*-type dopants if present at concentrations greater that ~1e16 atoms/cm^3.

STM is capable of providing electron density maps (from which atomic distribution maps can be derived) to atomic-scale spatial resolution (<0.3 nm). The sample may be a conductor or a semiconductor. Information is derived by scanning an atomically sharp tip over the surface of interest and measuring the current tunneling to/or from this and the sample. Measurement of *I-V* or *dI/dV-V* curves allows for electronic properties such as work function and electron density maps to be derived.

SOME REFLECTION/DIFFRACTION TECHNIQUES OF INTEREST

Another means of extracting chemical and/or structure information is by studying the reflection (specularly reflected) or diffraction (Thomson scattering) patterns produced on the elastic scattering of photons or electrons with solids. Specular reflection refers to the mirrorlike reflection observed off of smooth surfaces, that is, that in which deflection only occurs off of one plane (the surface plane). Diffraction, however, describes the coherent interaction of packages of photon or electron waves off specific crystalline surface planes, which may or may not be aligned with the surface plane.

More specifically, diffraction results from the interaction of photons or electrons with repeating lattice structures as described by the Bragg diffraction criteria (see Section 3.1.2.2). The diffraction is thus specific to the solid's lattice structure. This is particularly evident on crystalline substrates when the wavelength of the probe particles approaches atomic dimensions. For photons and electrons, this occurs at ~5000 and ~20 eV, respectively (defined through $E = hc/\lambda$). Since this occurs well below 1 eV for ions, this diffraction is not observed in the ion scattering methods (see Appendix F).

The techniques of greatest interest in this area (at least in the eyes of the author) include

X-ray Photoelectron Spectroscopy: An Introduction to Principles and Practices,
First Edition. Paul van der Heide.
© 2012 John Wiley & Sons, Inc. Published 2012 by John Wiley & Sons, Inc.

(a) X-ray diffraction (XRD)

(b) Glancing incidence diffraction (GID)

(c) X-ray reflectivity (XRR)

(d) Low-energy electron diffraction (LEED)

(e) Reflection high-energy electron diffraction (RHEED)

XRD is included even though it is not a surface-specific technique since this is by far the most common of the diffraction-based techniques used; that is, this is the standard method for solving crystal structures for both single-crystal samples as well as powdered crystalline samples. Surface specificity is lost in XRD due to the geometry used and the fact that photons have long path lengths within solids for both elastic and inelastic collisions. The remainder is surface-specific techniques.

H.1 XRD

This technique can be used to define the crystalline structure of any solid by directing an X-ray beam with a wavelength close to the separation of the atoms/ions making up the lattice and recording the diffraction pattern produced in the elastically scattered X-ray signal (those suffering Thomson scattering). At specific angles (that satisfying the Bragg diffraction criteria in Section 3.1.2.2), signals are produced that are specific to the long-range lattice structure. Thus, by rotating the sample/detector in 3-D, a full description of the lattice structure can be attained. This can also be applied to polycrystalline materials and powders, with average grain size data also supplied. Note: This is not a surface-specific technique.

H.2 GID

This surface-specific variant of XRD is used to define the crystallinity within surface layers from 10 to 200 nm in thickness. This does so by directing X-rays at glancing angles (~2°) with respect to the sample surface and by measuring the diffraction pattern produced in the non-specular beam (that deflected away from the plane normal to the surface that lies parallel to the incoming beam) when passed through a Soller slit arrangement. Analysis is carried out by holding the incident angle constant and by rotating the detector through some angular range.

H.3 XRR

Like GID, this technique directs an X-ray beam at the surface at glancing angles. However, instead of measuring the diffraction pattern, the specularly reflected beam is measured. This is done since X-rays undergo total external reflection at small grazing angles, with the resulting signal providing insight into the film thickness, roughness, and density. Both crystalline and amorphous films can be examined.

H.4 LEED

This electron analogue of XRD can provide the crystal structure over the outer 2–5 nm of a solids surface. This surface specificity arises from the fact that unlike photons, low-energy electrons (20–200 eV) can only travel a very short distance within a solid before suffering elastic and inelastic collisions (see Section 4.2.2.1). Electrons are directed normal to the surface and the backscattered signal (elastically scattered) is recorded. This displays an XRD-like diffraction pattern. Since electrons are used, HV and clean surfaces are required.

H.5 RHEED

This electron analogue of GID can provide the crystal structure of the outer 1–10 nm by directing a high-energy electron beam (~10–100 keV) at some glancing angle and measuring the forward scattered signal. The lattice order is revealed in the nonspecularly reflected beam as a result of the diffraction induced by the surface crystalline planes. Note: Although LEED provides better diffraction patterns, RHEED allows for improved sample access (needed in epitaxial growth studies). HV and clean surfaces are required.

TECHNIQUE ABBREVIATIONS LIST

AES	Auger electron spectroscopy
AFM	Atomic force microscopy
AR-XPS	Angle-resolved X-ray photoelectron spectroscopy
ATR	Attenuated total reflection
CAFM	Conductance atomic force microscopy
CSAFM	Current sensing atomic force microscopy (also called CAFM)
DRIFTS	Diffuse reflectance infrared spectroscopy
EBSD	Electron backscattered diffraction
EDS	Energy-dispersive X-ray spectroscopy (also called EDX)
EELS	Electron energy loss spectroscopy
EFM	Electrical force microscopy
EPMA	Electron probe microanalysis
ESCA	Electron spectroscopy for chemical analysis (also called XPS)
FIB	Focused ion beam
FIM	Field ionization microscopy

X-ray Photoelectron Spectroscopy: An Introduction to Principles and Practices,
First Edition. Paul van der Heide.
© 2012 John Wiley & Sons, Inc. Published 2012 by John Wiley & Sons, Inc.

FTIR	Fourier transform infrared spectroscopy
GID	Glancing incidence diffraction
HREELS	High-resolution electron energy loss spectroscopy
ICISS	Impact collision ion scattering spectroscopy
IR	Infrared
ISS	Ion scattering spectroscopy
KFM	Kelvin force microscopy
LATAP	Laser-assisted tomographic atom probe
LEED	Low-energy electron diffraction
LEEM	Low-energy electron microscopy
LEIS	Low-energy ion scattering
LEXES	Low-energy X-ray emission spectroscopy
MEIS	Medium-energy ion scattering
MFM	Magnetic force microscopy
NSOM	Near-field scanning optical microscopy (also called SNOM)
PEEM	Photoelectron emission spectroscopy
RAIRS	Reflection–adsorption infrared spectroscopy
RBS	Rutherford backscattering
REELS	Reflected electron energy loss spectroscopy
RHEED	Reflection high-energy electron diffraction
SAM	Scanning Auger microscopy
SARIS	Scattering and recoiling imaging spectroscopy
SCM	Scanning capacitance microscopy
SEIRS	Surface-enhanced infrared spectroscopy
SERS	Surface-enhanced Raman spectroscopy
SIMS	Secondary ion mass spectrometry
SNOM	Scanning near-field optical microscopy
SPELEEM	Spectroscopic photoemission and low-energy electron microscopy
SPM	Scanning probe microscopy
SSRM	Scanning spreading resistance microscopy
STEM	Scanning transmission electron microscopy
STM	Scanning tunneling microscopy
TAP	Tomographic atom probe
TEM	Transmission electron microscopy
TERS	Tip-enhanced Raman spectroscopy

TOF-SARS	Time-of-flight scattering and recoiling spectroscopy
TUNA	Tunneling atomic force microscopy
TXRF	Total reflection X-ray fluorescence
UPS	Ultraviolet photoelectron spectroscopy
WDX	Wavelength-dispersive X-ray analysis
XPEEM	X-ray photoelectron emission microscopy
XPS	X-ray photoelectron spectroscopy (also called ESCA)
XRD	X-ray diffraction
XRF	X-ray fluorescence
XRR	X-ray reflectivity

INSTRUMENT-BASED ABBREVIATIONS

AR-XPS	Angle-resolved X-ray photoelectron spectroscopy
CAE	Constant analyzer energy
CHA	Concentric hemispherical analyzer
CMA	Cylindrical mirror analyzer
CRR	Constant retard ratio
DLD	Delay line detector
DP	Depth profile
EM	Electron multiplier
FAT	Fixed analyzer transmission
FRR	Fixed retard ratio
HV	High vacuum
LV	Low vacuum
MCP	Microchannel plate
PAR-XPS	Parallel angle-resolved X-ray photoelectron spectroscopy
RDP	Relative depth plots
UHV	Ultrahigh vacuum

X-ray Photoelectron Spectroscopy: An Introduction to Principles and Practices,
First Edition. Paul van der Heide.
© 2012 John Wiley & Sons, Inc. Published 2012 by John Wiley & Sons, Inc.

GLOSSARY OF TERMS

Ab initio calculations   Calculations based on first principles (theory) and constants alone (cf. semiempirical and empirical calculations).

Adventitious carbon   The carbon layer forming on surfaces during analysis. This primarily takes the form of aliphatic hydrocarbons.

Angular asymmetry factor   Intensity variations in the recorded signals due to the X-ray source to analyzer angle (although generally fixed, this is instrument dependent).

Attenuation length (AL)   The average distance traveled by an electron of a specific energy within a particular multilayered and/or heterogeneous solid (may be amorphous or crystalline) between two successive inelastic scattering events, with elastic scattering included.

Auger process   De-excitation process resulting from the production of core holes. De-excitation results in the filling of the core hole by an electron closer to the Fermi edge. The energy difference is then removed via either an emission of an Auger electron or through fluorescence.

Background signal   The nonzero signal noted in spectra around peaks of interest. This arises from inelastic scattering of electrons.

X-ray Photoelectron Spectroscopy: An Introduction to Principles and Practices, First Edition. Paul van der Heide.
© 2012 John Wiley & Sons, Inc. Published 2012 by John Wiley & Sons, Inc.

Bandgap The gap between the upper edge of the valence band and the lower edge of the conduction band.

Binding energy (B.E.) Energy by which an electron is attracted to its nucleus. Note: That measured in XPS differs slightly from that in an unperturbed atom. These are differentiated as B.E. and B.E.$_{XPS}$, respectively.

Bremsstrahlung radiation Radiation in the form of photons that results when electrons are decelerated and/or deflected.

Charge compensation A method used to remove localized electrical charge buildup.

Charge potential model An electron density-based argument used to illustrate the cause of binding energy shifts due to initial state effects.

Chromatic aberration A type of optical distortion as described in Appendix E.

Collective oscillations Coherent vibration of a large number of particles (electrons and atoms) in a solid.

Collision cascade Ion impact-induced collision sequences in which momentum transfer occurs.

Core electron An atom or ion's inner electrons (cf. valence electrons).

Core hole That remaining after a core electron has been removed.

Curve fitting A method in which spectral features are fitting to some selection of peaks. Note: This is not a true deconvolution method since numerous solutions are possible.

Depth resolution Depth over which a signal from some abruptly appearing layer climbs from 16% of its maximum intensity to 84% (two standard deviations).

Diffraction A process resulting from the interaction of waves with each other. These can interfere constructively (results in a signal) or destructively (no signal).

Diffusion Mixing of atoms in a solid through some physical forces.

Elastic scattering Scattering process in which trajectories are altered but energy is not lost.

Electromagnetic force Electrostatic and magnetic force.

Electromagnetic radiation Oscillating electric and magnetic fields traveling through space.

Electromagnetic spectrum The complete range of wavelengths (frequencies) of electromagnetic radiation.

Electronegativity An arbitary scale developed to portray the reactivity of the elements.

Empirical   Experimental results (no theory used).

Escape depth   The distance normal to the surface from which 61.7% of the original photoelectron population originates. It is defined as $\lambda_{IMFP}\cos\Theta$.

Fermi edge (E_F)   Energy at which half the electronic orbitals in a solid are filled when at equilibrium. This is sometimes referred to as zero on the energy scale.

Final state effect   Effects stemming from the perturbation of the electronic structure, that is, that induced on core hole production.

Fluorescence   Emission of photons resulting from the filling of a core hole by an electron closer to the Fermi edge.

FWHM   Full width at half maxima (used to relay peak widths).

Gaussian distribution   A type of statistical function as described in Appendix D.

Holography   A method of extracting images from waves.

Inelastic mean free path   The average distance traveled by an electron of a specific energy within a particular single-layered homogeneous amorphous solid between two successive inelastic scattering events. This is represented as λ_{IMFP}.

Inelastic scattering   Scattering in which energy is lost (transferred to electronic excitation). Trajectories can remain unaltered.

Information depth   The depth below the surface from which a specified percentage of the photoelectrons emanate.

Initial state effect   Effects stemming from the unperturbed electronic structure.

Interatomic   Process between neighboring atoms/ions of interest.

Intra-atomic   Processes within the atom/ion of interest.

Ionization potential (I)   Minimum energy needed to remove an electron from a free atom/ion.

Kinetic energy ($K.E.$)   Energy contained within an electron that is measured.

Linear routine   A linear background subtraction routine.

Lorentzian distribution   A type of statistical function as described in Appendix D.

Maximum entropy   A deconvolution algorithm that uses Bayesian inference to satisfy the third law of thermodynamics (maximize entropy).

Microscopy   Any technique providing structural information.

Monochromatic   A stream of particles with a very narrow energy spread.

Multiplet splitting   The splitting in peaks from any level in ions with unpaired valence electrons. This is due to the spin interactions of valence and unpaired core electrons.

Pass energy   The energy of electrons as they pass through a concentric hemispherical analyzer (CHA) energy filter (they may be accelerated/decelerated to this energy).

Passivating oxide   Thin stable protective oxide.

Phonons   Collective vibrations of atoms within a solid.

Photoelectron cross section   Probability describing the likelihood for photoelectron production from a specific core level of a specific element under irradiation of photons of a specific energy. Scofield values are used in XPS.

Photoelectrons   X-ray-induced electron emission.

Photons   Electromagnetic radiation (a package of energy with zero rest mass).

Plasmon loss features   Repeating peaks (in units of energy) noted on the higher binding energy side of a photoelectron peak in XPS spectra due to plasmon formation.

Plasmons   Collective vibrations of electrons within a solid.

Poisson distribution   A type of statistical function as described in Appendix D.

Principal component analysis   A mathematical procedure that attempts to recognize the minimum number of variables (principal components) in a data set.

Quantum mechanics   A mathematical interpretation of an atomic structure.

Quantum numbers   A scheme based on quantum mechanics in which the energy, momenta, spatial distribution, and spin of each electron bound to a specific element are specified.

Sample rotation   Rotation of the sample around the area being analyzed with the axis of rotation being perpendicular to the surface.

Sampling depth   Depth from which 95.7% of all photoelectrons emanate (equivalent to $3\lambda_{IMFP} \cos\Theta$).

Segregation   Demixing of atoms within a solid as a result of chemical forces.

Semiempirical   Calculations in which known parameters, that is, structure, are input.

Sensitivity   Minimum concentration for which a particular element can be detected.

Shake-off process Excitation of valence electrons to unoccupied unbound states (these rarely result in peaks).

Shake-up satellites Additional peaks noted at a higher binding energy relative to a peak of interest arising from the excitation of valence electrons to unoccupied bound states.

Shirley routine A background subtraction routine devised by Shirley.

Siegbahn notation A notation scheme used to describe X-ray emissions.

Space charge effects Electrostatic repulsion of electrons or ions in a charged particle beam.

Spatial resolution Distance over which a signal from some abruptly appearing interface climbs from 16% of its maximum intensity to 84% (two standard deviations).

Specificity In XPS, this means to examine a specific region, that is, the surface region.

Spectrometry Any technique providing chemical information (difference from spectroscopy is discussed in Appendix F).

Spectroscopic notation A notation scheme used to describe stationary states.

Spectroscopy Any technique providing chemical information (difference from spectrometry is discussed in Appendix F).

Spherical aberration Type of optical distortion as described in Appendix E.

Spin orbit splitting The splitting in peaks from levels (stationary states) of nonzero l (angular momentum quantum number) due to the spin interactions of protons with electrons.

Sputter rate Rate in units of depth over time at which the surface is being removed through sputtering.

Sputter yield Number of surface atoms removed per incoming ion used to sputter the surface.

Sputtering The removal of atoms from a solid surface as a result of energetic ion impact.

Stationary states Specific energy levels in which bound electrons can reside.

Surface The outermost region of any solid or liquid substrate that dictates how the solid interacts with its environment.

Takeoff angle The angle between the sample surface and the axis of the electron collection optics.

Tougaard routine   A background subtraction routine devised by Tougaard.

TPP-2M relation   Semiempirical expression developed to approximate the value of λ_{IMFP}.

Uncertainty principle   Otherwise known as the Heisenberg uncertainty principle. This states that both the momentum and position of any particle cannot be simultaneously defined to any degree of precision (only one or the other can be accurately defined).

Vacuum   Any gaseous environment in which the density of particles exerts a lesser pressure than its gaseous surroundings pressure.

Vacuum energy (E_{vac})   The true zero on the energy scale. This is the energy that defines the border between free electrons in the gas phase and electrons bound to an atom, molecule, or solid.

Valence electron   An atom or ion's outer electrons (cf. core electrons).

Work function   The minimum energy required to remove an electron from a solid surface (the difference in energy between the vacuum level and Fermi energy).

X-ray notation   A notation scheme used to describe stationary states.

Z + 1 model   A model used to approximate core-level binding energies.

Zalar rotation   Trademarked name introduced by physical electronics to describe sample rotation.

QUESTIONS AND ANSWERS

(1) For a ground-state neutral atom with 13 protons, describe
- **(a)** Which element this is
- **(b)** The quantum numbers, n, and l of the inner two core electrons
- **(c)** The stationary state these inner two core electrons reside in using both spectroscopic and X-ray notation
- **(d)** The $B.E._{XPS}$ of photoelectrons emitted from the $2p_{3/2}$ level of this atom when present in its elemental solid form (refer to Appendix B for tabulated values)

Answers:
- **(a)** Al
- **(b)** $n = 1, l = 0$
- **(c)** 1s, K
- **(d)** 72.55 eV

(2) A neutral ground-state atom containing 29 electrons present within its elemental solid with a work function of 4.5 eV experiences

X-ray Photoelectron Spectroscopy: An Introduction to Principles and Practices,
First Edition. Paul van der Heide.
© 2012 John Wiley & Sons, Inc. Published 2012 by John Wiley & Sons, Inc.

photoelectron emission upon Al-$K\alpha$ X-ray irradiation (1486.6 eV). The most intense photoelectron emissions have a $K.E._{XPS}$ energy of 549.7 eV. Describe

(a) Which element the emissions are coming from
(b) The $B.E._{XPS}$ of the respective photoelectrons
(c) The stationary state from which these emanated using spectroscopic notation and also X-ray notation
(d) The subsequent L_3V transition (results in Auger emissions or fluorescence) using spectroscopic notation

Answers:
(a) Cu
(b) 932.7 eV
(c) $2p_{3/2}$, L_3
(d) $3d \rightarrow 2p_{3/2}$

(3) Al $K\alpha_1$ X-rays (Siegbahn notation) arise from KL_3 transitions (X-ray notation). In spectroscopic notation, this fluorescence results from the transition of an electron from the $2p_{3/2}$ level to fill a core hole in the 1s level. The fluorescence energy is thus the energy difference between these states minus some small perturbation energy (final state effects). Assuming no perturbation occurs, what is the energy of the $K\alpha_1$ X-rays from (a) Mg and (b) Si? Use values listed in Appendix B.

Answers:
(a) 1253.5 eV
(b) 1739.58 eV

(4) If you were given a choice between using the Al or Ag anodes in a monochromatic source (from which the Al-$K\alpha$ or Ag-$L\alpha$ X-rays are used), which would you use if you were interested in ascertaining the $B.E._{XPS}$ value of the Si-1s core electrons from Si?

Answer:
The Ag anode would be used since only the Ag-$L\alpha$ X-rays have the energy needed to induce the emission of Si-1s core electrons (see Table 3.3).

(5) Describe if and how the work function of a sample of interest can be derived using X-ray photoelectron spectroscopy (XPS) or ultraviolet photoelectron spectroscopy (UPS).

Answer:
Assuming sufficiently high-energy resolution conditions are used, the work function of the sample being analyzed can be defined from the low $K.E._{XPS}$ onset (high $B.E._{XPS}$ drop-off) of the spectra collected normal to the sample surface, that is, the high $B.E._{XPS}$ cutoff that appears close to the X-ray energy being used to generate the photoelectrons. The work function should equate to the energy difference between the X-rays used and the $B.E._{XPS}$ value where the signal drops to 50% of its peak value (this is due to the finite energy spread of the photoelectrons recorded in the respective XPS instruments).

Note: UPS using a discharge lamp can provide the work function values to a far superior precision than XPS using a monochromated source due to the narrower line width of discharge lamps relative to monochromated sources.

(6) In typical XPS spectra, both photoelectron peaks and Auger electron peaks are present. The questions are
  (a) What instrumental parameter can be used to distinguish the two?
  (b) What is the result?

Answers:
  (a) Use different X-ray source energies and compare the spectra.
  (b) The Auger peaks move to different $B.E._{XPS}$ values as the source energy is changed, whereas the photoelectron peaks remain at the same $B.E._{XPS}$.

(7) Describe what happens to all the photoelectron peaks observed in the spectra collected from an insulating sample when ineffective charge neutralization conditions are used.

Answer:
All detectable photoelectron peaks from all levels and from all elements present within the analyzed volume move by the same amount to higher $B.E._{XPS}$ values. In extreme cases, these peaks will become part of the background signal.

(8) A compound contains CH-, CO-, and CF-based organics. Using the knowledge that the $B.E._{XPS}$ shifts for the C-1s level can be ascribed to typical initial state effects (intra-atomic), list these in the expected $B.E._{XPS}$ order from lowest to highest.

Answer:

CH < CO < CF. This is understood since the electronegativities of these elements scale in the same fashion.

(9) What unusual initial state effect causes core-level $B.E._{XPS}$ values of certain metal atoms/ions to decrease during oxidation? Note: Typically, the oxidation of metals results in an increase in their core-level $B.E._{XPS}$ values. This unusual effect is, however, observed during the oxidation of select metals, including Sr and Ba.

Answer:

This unusual effect occurs when interatomic initial state effects dominate over intra-atomic effects (typically, this is the other way around). Note: Both have apposing effects on the direction of the $\Delta B.E._{XPS}$ during the oxidation of the photoelectron emitting atom/ion.

(10) What could cause the spin orbit splitting value of photoelectrons from a specific level or a specific element to deviate from that typically observed?

Answer:

If the element also suffers multiplet splitting during photoelectron emission, this can cause the apparent spin orbit splitting energy to change.

(11) From where do the increased final state effects in the form of rearrangement as well as the requirement to use Russell–Saunders coupling arguments (L-S) in describing spin orbit splitting for the higher Z elements stem?

Answer:

Increased coupling that occurs between electrons in different stationary states. In the light elements, electrons in different stationary states can be considered as acting independently; that is, l and s (m_s) do not interact.

(12) Calculate the theoretical bulk plasmon energy resulting from photoelectron emission from Cs present within its elemental solid (mass density = 1.90 g/cm^3).

Answer:

(a) 3.33 eV

(13) What is the flaw in Koopman's theorem when applied to examining XPS spectra?

Answer:
A frozen orbital approach is used; that is, this assumes no rearrangement/relaxation occurs for the valence electrons of the photoelectron emitting atom/ion.

(14) The analysis of an unknown but homogeneous sample is carried out with photoelectrons collected at some off-normal takeoff angle. On rotating the sample around its normal axis (azimuth rotation), reproducible periodic spikes are noted in various core-level photoelectron signals. The questions are
(a) What are these variations indicative of?
(b) How may they be useful?

Answers:
(a) The variations in photoelectron intensities are due to photo-electron diffraction induced by the crystal the electrons emanate from (the sample)
(b) This will only occur on a single-crystal substrate. Photoelectron diffraction can be useful in revealing the element-specific surface crystallographic structure.

(15) What is the difference between a CHA and a spherical mirror analyzer (SMA)?

Answer:
Both are types of energy filters and both direct the electrons of interest between two hemispherical plates held at specific potentials. The difference lies in the fact that the average radii of the hemispherical plates used in a CHA are equal to the trajectory of the desired electrons. In an SMA, the radii of the desired electrons are much smaller.

Note: SMAs are presently only used in XPS in parallel imaging studies.

(16) Can XPS be used to define film thickness?

Answer:
If the film is less than the sampling depth and some spectral difference can be discerned between the film and the underlying

substrate, then the answer is yes. Indeed, very precise measurements can be carried out on uniform films through a modification of the Beer–Lambert Law.

(17) Quantification of sputter depth profiles of even the most stable metal oxides (e.g., perovskites) with 0.5-keV Ar^+ ions at an angle of 45° relative to the sample surface, using sensitivity factors defined from unsputtered surfaces, tends to reveal less than stroichiometric amounts of O. Explain why this occurs.

Answer:
Preferential removal of light elements relative to heavy elements typically occurs during sputtering of oxides, nitrides, and so on. This occurs as a result of the mass-dependent momentum transfer occurring between the incoming ion (Ar^+) and the element it interacts with, which in effect alters the sputter yields of the respective elements even though sputtered under the same conditions. This effect, termed preferential sputtering, can only be accounted for if the sensitivity factors are adjusted accordingly during quantification. Note: Sputter-induced elemental migration (diffusion, segregation, etc.) can also occur.

(18) Scanning electron microscopy (SEM), which records secondary electrons (these peak in energy within the 1- to 10-eV range), is often used in conjunction with Auger electron spectroscopy (AES) and/or XPS to visualize the area of interest (this is done since SEM provides the best spatial resolution of the three, even though it does not provide elemental characterization capabilities). Define which of these three techniques exhibit the greatest surface specificity to the formation of carbon overlayers (this could be in the form of adventitious C, self-assembled monolayers, amorphous carbon layers, etc.) on gold. Assume that in all cases, the data are collected at normal takeoff angles, C-1s photoelectron emissions resulting from Al-$K\alpha$ irradiation are examined in XPS, and C KVV Auger emissions are examined in AES.

Answer:
AES will be the most surface specific to the formation of carbon overlayers on gold. This is realized since the inelastic mean free path of the C KVV Auger electrons is the shortest. Indeed, the surface specificity is a function of the electron inelastic mean free path. This exhibits a strong dependence on the electrons' kinetic

energy (*K.E.*), with a minima in the inelastic mean free path noted at a *K.E.* of ~80 eV for gold. Values then steadily increase with increasing *K.E.* above this minima and increase more rapidly with decreasing *K.E.* below this minima (see Fig. 4.6). The C *KVV* (*KLL*) energy can be approximated as the difference between the C-1s and valence levels minus the instruments' work function (typically in the 3- to 5-eV range). The *K.E.* of the C-1s photo-electrons is well in excess of this (>1200 eV) and, as stated in the question, the *K.E.* of the secondary electrons is well below this (these peak at between 1 and 10 eV).

Note: The reduced surface specificity of SEM also explains why this technique does not require the stringent vacuum conditions needed in AES as well as XPS; that is, SEM essentially sees through any adventitious C layers formed. The improved surface specificity of AES, however, explains why this technique is commonly used in examining such things as the surface contamination level during epitaxial growth.

XPS VENDORS

JEOL	JEOL USA Inc. 11 Dearborn Road Peabody MA 01960 USA http://www.jeol.com	Complete instruments
Kratos	Kratos Analytical Wharfside Trafford Wharf Road Manchester United Kingdom http://www.kratos.com	Complete instruments
Omicron	Omicron Vakuumphysik GmBH Idstreiner Straße 78 65232 Taunusstein Germany http://www.omicron- instruments.com	Complete instruments Specialty parts

X-ray Photoelectron Spectroscopy: An Introduction to Principles and Practices,
First Edition. Paul van der Heide.
© 2012 John Wiley & Sons, Inc. Published 2012 by John Wiley & Sons, Inc.

Phi	Physical Electronics Inc. 18725 Lake East Drive Chanhassen MN 55317 USA http://www.phi.com	Complete instruments
Revera	ReVera Incorporated 3090 Oakmead Village Drive Santa Clara CA 95051 USA http://www.revera.com	Complete instruments (inline fab based)
Specs	Specs GMBH Voltastraße 5 13355 Berlin Germany http://www.specs.de	Specialty parts
Staib	Staib Instruments GmBH Obere Haupstraße 45 85354 Freising Germany http://www.staib-instruments.com	Specialty parts
Thermo-Fisher	The Birches Industrial Park Imberhorne Lane East Grinstead West Sussex RH19 1UB United Kingdom http://www.thermoscientific.com	Complete instruments
VG Scienta	Gammadata Scienta P.O. 15120 Vallongatan 1 750 15 Uppsala Sweden http://www.gammadata.se	Specialty parts

VSW	VSW Ltd.	Specialty
	Unit 4, Heater Close	parts
	Lyme Green Business Park	
	Macclesfield, Cheshire, SK110LR	
	United Kingdom	
	http://www.vsw.co.uk	

Refurbished systems

RBD Instruments	RBD Instruments	Complete
	2437 NE Twin Knolls Drive	instruments
	Suite 2,	Specialty
	Bend OR, 97701	parts
	USA	
	http://www.rbdinstruments.com	

REFERENCES

Auger P, *J. Phys. Radium.*, **6**, 205 (1925).

Bagus PS, Illas F, Pacchioni G, Parmigiani F, *J. Electron. Spectrosc. Relat. Phenom.*, **100**, 215 (1999).

Barr TL, Liu YL, *J. Phys. Chem. Solid*, **7**, 657 (1989).

Bearden JA, Burr AF, *Rev. Mod. Phys.*, **39**, 125 (1967).

Berglund CN, Spicer WE, *Phys. Rev. A*, **136**, 1030 and 1044 (1964).

Bethe H, *Annalen der Physik*, **397**, 325 (1930).

Born M, *Verh. Dt. Phys. Ges.*, **21**, 13 and 679 (1919).

Bouguer P, *Essai d'Optique sur la Gradation de la Lumiere*, Paris (1729).

Bragg WL, *Proc. Camb. Philol. Soc.*, **17**, 43 (1913).

Briggs D, Seah MP, *Practical Surface Analysis, Volume 1: Auger and X-ray Photoelectron Spectroscopy*, 2nd Edition, John Wiley & Sons, New York (1990).

Cardona M, Ley L, *Photoemission in Solids I: General Principles*. Springer, Berlin (1978).

Chadwick J, *F.R.S. Proc. Roy. Soc. A*, **136**, 692–708 (May 1932).

Chong DP, *Chem. Phys. Lett.*, **232**, 486 (1995). *J. Chem. Phys.*, **103**, 1842 (1995).

Citrin PH, Thomas TD, *J. Chem. Phys.*, **57**, 4442 (1972).

X-ray Photoelectron Spectroscopy: An Introduction to Principles and Practices, First Edition. Paul van der Heide.
© 2012 John Wiley & Sons, Inc. Published 2012 by John Wiley & Sons, Inc.

Cole RJ, Macdonald BF, Weightman P, *J. Electron. Spectrosc. Relat. Phenom.*, **125**, 147 (2002).

Condon EU, Shortley GH, *The Theory of Atomic Spectra.* Cambridge University Press, Cambridge (1935).

Doniach S, Sunjic M, *J. Phys. C*, **3**, 285 (1970).

Einstein A, *Ann. Phys.*, **17**, 549 (1905).

Eisberg R, Resnick R, *Quantum physics of Atoms, Molecules, Solids, Nuclei and Particles*, 2nd Edition, John Wiley & Sons, New York (1985).

Fadley CS, Hagstroem SBM, Hollander JM, Klein MP, Shirley DA, *J. Chem. Phys.*, **48**, 3779 (1968).

Fuggle JC, Mårtensson N, *J. Electron. Spectrosc. Relat. Phenom.*, **21**, 275 (1980).

Gaarenstroom SW, Winograd NJ, *Chem. Phys.*, **67**, 3500 (1977).

Gelius U, *Phys. Scr.*, **2**, 70 (1970).

Gilbert RA, Llewellyn JA, Swartz WE Jr., Palmer JW, *Appl. Spectrosc.*, **36**, 428 (1982).

Gnaser H, *Low Energy Ion Irradiation of Solid Surfaces.* Spinger, Berlin (1999).

Griffiths DJ, *Introduction to Quantum Mechanics*, 2nd Edition, Prentice Hall, New York (2004).

Harber F, *Verh. Dt. Phys. Ges.*, **21**, 750 (1919).

Hertz H, *Annalen der Physik*, **33**, 983–1000 (1887).

Herzog W, Schwarz M, Sixl H, Hoppe R, *Z. Phys. B Condens. Matter*, **71**, 19 (1988).

Hubbard J, *Proc. Roy. Soc. A* **176**, 328 (1963).

Hüfner S, *Photoelectron Spectroscopy: Principles and Applications*, 3rd Edition, Springer, Berlin (2003).

Ibach H, in *Electron Spectroscopy for Surface Analysis*, ed. Ibach H, Topics, Curr. Phys., Vol. 4, Berlin (1977).

Imada M, Fujimori A, Tokura Y, *Rev. Mod. Phys.*, **70**, 1039 (1998).

Jolly WL, in *Electron Spectroscopy: Theory, Rechniques and Applications I*, eds Brundle CR, Baker AD, Academic Press, New York (1977).

Jolly WL, Hendrickson DN, *J. Am. Soc.*, **92**, 1863 (1970).

Jona F, Shirane G, *Ferroelectric Crystal.* Dover, New York (1993).

Kissell KR, Preparation of iodine SWNTs and iodine US-tubes: synthesis and spectroscopic characterization of iodine-loaded SWNTs for computed tomography molecular imaging, Ph.D. Thesis, Rice University (2006).

Kissell KR, Hartman KB, van der Heide PAW, Wilson LJ, *J. Phys. Chem.*, **110**, 17425 (2006).

Koopmans T, *Physica*, **1**, 104 (1933).

Kotani A, Toyozawa Y, *J. Phys. Soc. Jpn.*, **37**, 912 (1974).

Magee CA, *Nucl. Instru. Meth.*, **191**, 297 (1981).

Mahoney CM, *Mass Spectrometry Reviews*. DOI 10.1002/mas.20233 (http://www.interscience.wiley.com) (2009).

Meitner L, Über die Entstehung der β-Strahl-Spektren radioaktiver Substanzen. *Z. Physik*, **9**, 131 (1922).

Mitchell DF, Clark KB, Bardwell JA, Lennard WN, Massoumi GR, Mitchell IV, *Surf. Interface Anal.*, **21**, 44 (1994).

Moretti G, *J. Electron. Spectrosc. Relat. Phenom.*, **95**, 95 (1998).

Mott NF, *Proc. Phys. Soc. London Sect. A* **62**, 416 (1949).

Moulder JF, Sickle WF, Sobol PE, Bomben KD, *Handbook of X-ray Photoelectron Spectrosdcopy*, Physical Electronics Inc., Eden Prairie (1992).

Nagoshi M, Syono Y, Tachiki M, Fukuda Y, *Phys. Rev. B*, **51**, 9352 (1995).

Pollak RA, Lee L, McFeely FR, Kowalczyk SP, Shirley DA, *J. Electron. Spectrosc. Relat. Phenom.*, **3**, 38 (1974).

Purcell EM, *Phys. Rev.*, **54**, 818 (1938).

Redhead PA, Hobson JP, Kornelsen EV, *The Physical Basis of UltraHigh Vacuum*, American Institute of Physics, New York (1993).

Reilman RF, Msezane A, Manson ST, *J. Electron. Spectrosc. Relat. Phenom.*, **8**, 389 (1976).

Renault O, Lavayssiere M, Bailly A, Mariolle D, Barrett N, *J. Electron Spectrosc. Related Phenom.* **171**, 68 (2009).

Ritchie RH, *Phys. Rev.*, **106**, 874 (1957).

Röntgen WC, (1985) Translated by Stanton, Arthur. *Nature*, **53**, 274 (1896).

Roscoe HE, *John Dalton and the Rise of Modern Chemistry*, Century science series. Macmillan, New York. Retrieved 2011-04-03 (1895).

Rutherford E, *Philos. Mag*, **21**, 669–688 (April 1911).

Scofield JH, *J. Electron. Spectrosc. Relat. Phenom.*, **8**, 129–137 (1976).

Seah MP, Dench WA, *Surf. Interface Anal.*, **1**, 2 (1979).

Seo HW, *Evolution of Structural Perfections in High T_C Superconducting Thin Films*, PhD Thesis, University of Houston (2005).

Shirley DA, *Chem. Phys. Lett.*, **15**, 325 (1972a).

Shirley DA, *Phys. Rev. B*, **5**, 4709 (1972b).

Shirley DA, *Adv. Chem. Phys.*, **23**, 85 (1973).

Shirley DA, Fadley CS, *J. Electron. Spectrosc. Relat. Phenom.*, **137-140**, 43 (2004).

Siegbahn K, ESCA-Atomic, Molecular and Solid State Structure Studied by Means of Electron Spectroscopy (1967).

Sigmund P, *Phys. Rev.*, **184**, 383 (1969).

Sigmund P, in *Sputtering by Ion Bombardment III*, ed. Behrisch R, Springer, Berlin (1981).

Simonsen AC, Yubero F, Tougaard S, *Surf. Sci.*, **436**, 149 (1999).

Skinner HWB, *Proc. Roy. Soc. London*, **A135**, 84 (1932).

Smalley RE, *J. Vac. Sci. Technol. A*, **19**, 1800 (2001).

Sosulnikov MI, Teterin YA, *J. Electron. Spectrosc. Relat. Phenom.*, **59**, 111 (1992).

Stokes GG, *Phil. Trans. Roy. Soc. London*, **142**, 463 (1852).

Tanuma S, Powell CJ, Penn OR, *Surf. Interface Anal.*, **21**, 165 (1993).

Teterin YA, Sosulnikov MI, Petrov YA, *J. Electron. Spectrosc. Relat. Phenom.*, **95**, 95 (1998).

Thomas TD, *J. Am. Soc.*, **92**, 4181 (1970).

Thomson JJ, *Philos. Mag*, **44**, 293 (1897).

Tougaard S, *Surf. Interface Anal.*, **11**, 453 (1988).

van der Heide PAW, *Surf. Interface Anal.*, **33**, 414 (2002).

van der Heide PAW, unpublished data (2005).

van der Heide PAW, *J. Electron. Spectrosc. Relat. Phenom.*, **151**, 79 (2006).

van der Heide PAW, *J. Electron. Spectrosc. Relat. Phenom.*, **164**, 8 (2008).

van Vleck JH, *Phys. Rev.*, **45**, 405 (1934).

Vasquez RP, *J. Electron. Spectrosc. Relat. Phenom.*, **66**, 209 (1994).

Vasquez RP, Foote MC, Bajuk LJ, Hunt BD, *J. Electron. Spectrosc. Relat. Phenom.*, **57**, 317 (1991).

Vasquez RP, Hunt BD, Foote MC, Bajuk LJ, Olson WL, *Phys. C*, **190**, 249 (1992).

Wagner CD, *Analyt. Chem.*, **44**, 967 (1972).

Wagner CD, Gale LH, Raymond RH, *Analyt. Chem.*, **51**, 466 (1979).

Wagner CD, Naumkin AV, Kraut-Vass A, Allison JW, Powell CJ, Rumble JR Jr., *NIST Standard Reference Database 20*, Version 3.5, http://srdata.nist.gov/xps/ (2003).

Wertheim GK, *J. Electron. Spectrosc. Relat. Phenom.*, **34**, 303 (1984).

Williams GP, in *X-Ray Data Booklet, section 1.1: Electron Binding Energies*, ed. Thompson AC, Vaughan D, 2nd Edition, Lawrence Berkeley National Laboratory, University of California (2001). Also at http://xdb.lbl.gov

Zannan J, Sawastsky GA, Allen JW, *Phys. Rev. Lett.*, **55**, 418 (1985).

Zashkvara VV, Korsunskii MI, Kosmachev OS, *Sov. Phys. Tech. Phys.*, **11**, 818 (1966).

Zhang FC, Rice TM, *Phys. Rev. B*, **37**, 3759 (1988).

INDEX

Note: Page entries in **bold** indicate pages on which the main discussion of the topic occurs.

X-ray Photoelectron Spectroscopy: An Introduction to Principles and Practices,
First Edition. Paul van der Heide.
© 2012 John Wiley & Sons, Inc. Published 2012 by John Wiley & Sons, Inc.

Printed in the United States
By Bookmasters